A PARLIAMENT OF SCIENCE

A PARLIAMENT OF SCIENCE

Science for the 21st Century

Edited by

Michael Tobias
Teun Timmers
Gill Wright

State University of New York Press

For information, address State University of New York Press,
90 State Street, Suite 700, Albany, NY 12207

Production by Marilyn P. Semerad
Marketing by Anne M. Valentine

Library of Congress Cataloging-in-Publication Data

A parliament of science : science for the 21st century / edited by
Michael Tobias, Teun Timmers, Gill Wright
    p.   cm.
   Includes bibliographical references and index.
   ISBN 0–7914–5813–x (acid-free paper)—ISBN 0–7914–5814–8 (pbk.: acid-
free paper)
   1. Science—Forecasting. 2. Technological forecasting. 3. Twenty-first
century—Forecasts. I. Tobias, Michael. II. Timmers, Teun. III Wright, Gill.
   Q158.5 .P38 2003
   303.48'3'0905—dc21   2002036480

10 9 8 7 6 5 4 3 2 1

Contents

Acknowledgments

The editors gratefully acknowledge support from the International Council for Science (ICSU) and the United Nations Educational, Scientific, and Cultural Organization (UNESCO), especially Dr. Howard Moore, Secretary of the World Conference, and Mme Hélène-Marie Gosselin, Director of the Office of Information. The American Association for the Advancement of Science (AAAS) was very helpful during the 1999 Annual Symposium.

Sir Joseph Rotblat kindly gave permission to print his interview while retaining its copyright.

Andrew MacLean and Victoria Earle contributed significantly by doing research for this book. A special thanks also to Jane Gray Morrison and Kathryn P. Davison. Ms. Katalin Bogyay, Director of the Hungarian Cultural Centre in London, independent television producer and writer, was the producer during the filming of the interviews in Budapest and Anaheim. The editors are grateful for space generously provided by Hotel Flamenco in Budapest.

The editors are particularly indebted to the World Watch Institute for permission to quote extensively from *State of the World 1998* and *Vital Signs 1999*.

MICHAEL TOBIAS

Introduction

The Genesis of the Book

During January and June 1999, thousands of scientists and policymakers from over one hundred fifty countries gathered in Anaheim, California, and Budapest, Hungary, to examine the complex roles and interrelations of science, ethics, policy, environment, and technology for the twenty-first century.

These two end-of-the-century summits, which in combination comprised the largest, most diverse gatherings of scientists in history, were sponsored by the AAAS (American Academy for the Advancement of Science), UNESCO (United Nations Educational, Scientific, and Cultural Organization) and ICSU (International Council for Science).

Filmmakers Gill Wright, Teun Timmers, and Michael Tobias had been commissioned by UNESCO—with the support of the AAAS and ICSU—to film a multipart documentary series for television incorporating filmed interviews from these two important gatherings. Three documentary teams from the United States, the United Kingdom, and Hungary in all captured over sixty in-depth thirty- to ninety-minute sit-down discussions with individual scientists, the interviews conducted by Tobias and Timmers. In addition, major addresses were also filmed, as well as dozens of shorter "stand-up" interviews with other visiting scientific delegates, and an in-depth roundtable discussion with nearly twenty younger scientists from around the world.

The Style of the Book

Great thinkers are not always, of course, the best speakers or writers. In the case of these chosen scientists, participants have been selected on the basis of three qualities: their intellectual stature, their diversity of commitments, and their facilities for expression. The resulting styles have meant the accommodation of a wide range. Great ideas, brought forth under the "pressure" of live cameras, engendering a degree of candor that is highly unusual; a forthrightness that can claim no fallback to more studied, abstract methods of conveyance.

The book provides accessible platforms of open discussion and contemplation. The interview questions have been delegated to topic headings, as it were; deleted in practice to ensure a smoother, more compact, and readable presentation.

The subject matter ranges from North/South economics and the scientific brain drain; to biodiversity; sustainability; politics; cloning; life and death ethics; medicine in the twenty-first century; high-energy physics; the Big Bang; United States, European, and broad African science policies; the fate of the Earth; the role of education, the plight of women in the sciences; the role of lending agencies like the World Bank; communication in science; the ozone hole; the rain forests; coral ecology; the relationship between industry and science; AIDS; indigenous folklore as a form of critical watershed science; global warming; ethics and technology; sustainability and agriculture; evolutionary paleontology; animal rights; and prospects for human settlement on Mars.

Interviewees come from the United States, Israel, Ireland, the United Kingdom, Sweden, India, Hungary, Sudan, France, Spain, Brazil, Cuba, Canada, Denmark, Germany, Egypt, and Belgium.

The book serves as a fine guide to the issues and nuances of twenty-first century scientific thought. These are not "prepared" pieces for an anthology but a lively, often iconoclastic dialogue that is distinguished by its frankness and willingness to tackle hard questions; and by the power of the media to elicit a style of delivery that can be assimilated by readers of many ages and disciplines.

As such, the work should prove fascinating to all students of social, natural, and "hard" sciences; to teachers, parents, and those interested in public policy, futurism, and ethics. It is a rich overview of the state of the scientific world; its obsessions, fears, ghosts, and high ambitions. Nearly all of the pieces devolve, at some point, to the speaker's own personal life; her or his hopes, dreams, and enthusiasm. In this regard, *A Parliament of Science* is an important window on the roots of scientific discovery; what it is that compels individuals to embark on a career in science; what specific questions trigger scientific investigation, as opposed to some other avenue of discovery.

With its ethical focus throughout, and with the raising of hard questions that scientists, policy makers, and the public must address, *A Parliament of Science* is no simple celebration of the works of science but, rather, a sobering reappraisal of where we've been, what our ingenuity has wrought for better or worse, and where we and the whole planet seem to be headed. To contextualize science in this arena is new—given the range of interviewees—and should prove to be a great stimulus to added thought and discussions.

Historical Context

The ethical root of so much of the foregoing raises an historical perspective worth commenting upon here. Francis Bacon, Viscount St. Albans (1561–1626), is often credited with having been the first philosopher of science as presented in his work *Advancement of Learning*. But where Bacon failed to account for the disposition of facts, or for the ability of scientists to discriminate among the useful ideas of their peers and predecessors, was with regard to the whole arena of judgment. In Latin *scientia* (knowledge) connotes no other responsibility or moral calling. It is, strictly, the realm of facts, without obligation, application, or duty. Indeed, Bacon's own greatest work of science, *Novum Organum* (The New Instrument), was composed in Latin, a language known by only a minute coterie of the public. It thus confirmed a long-standing bias in favor of "experts," precluding the "common man" from gaining access to that realm of divine knowledge (read: divine grace). Knowledge has always been tantamount to power and privilege, both domains of which—in any society—imply relations between people that must—at the risk of great peril or prejudice—be moderated. Hence, from the time of Aristotle, there was no way to separate science from politics or ethics.

Bacon was conflicted over this dichotomy that pitted the real world, with its many tumbles and all too human nature, against the pure sphere of the fact. His proposals for scientific method laid the groundworks for an irreconcilability: Ethics and moral judgment beside the direct observation of nature with its Platonic Forms. From these two realms must necessarily come a perfect synthesis. But it scarcely exists.

A thousand years ago, it existed with even less likelihood than in Bacon's Renaissance days. On the momentous eve of a new millennium, A.D. 999, Gerbert de Aurillac, Pope Sylvester II, one of the most learned people in papal history, had to balance the fears of the multitudes across Europe—the battle between the biblical Gog and Magog, the Apocalypse, and Armageddon—with his far more sober, scientific learning. Sylvester had memorized Aristotle, Cicero, Porphyry, and Boethius. Taught poetry and logic, and mathematics and astronomy. Loved literature. Musician, author, philosopher, this uncommon Pope dispensed with most Biblical commentary and built his own globes to recreate the known planets; fashioned a sundial, and tinkered assiduously with an abacus. His library was grand. Imagine, then, his dilemma, upon being thought of as the mouthpiece for God, a possibly compromising, if embarrassing situation to begin with for a man of science; then having to weigh the future of the world, while tens of thousands of fearful, unlettered denizens of Rome stood all night in St. Peter's Square waiting for some divine embrace. Was redemption possible in a world of doubt? The stakes were ultimate. One man of science, wearing the highest robes of the Vatican, against the unknown.

Pope Sylvester chose wisely, cautioning his minions to fear not, to have faith. A very politically savvy strategy. Humanity saw the world carry on the day after, business more or less as usual. But consider what science was, in Sylvester's age: Arabic alchemy, only the first hint of the notion of a chemical laboratory, the "zero" only recently introduced to computational analysis, astrology still passing for astronomy in most people's minds, no clue about oxygen, and—except for the rare Avicenna or Rhazes (generalists with groundbreaking ideas about physiology and pharmacology)—the practice of medicine was more primitive than in Greek times, fourteen hundred years earlier, and life expectancy still hovered around thirty-two years, as it had during the time of Christ.

By the period of Francis Bacon, much had changed, but much had not. There had been remarkable strides in lens crafting, allowing unprecedented views into the galaxy, and into the insides of living organisms. Medicine, while still barbaric by even yesterday's standards, had, at least, replaced alchemy. In Japan the first anesthetics were coming into being. Astrology was gone. Earth sciences were competing with Biblical conceptions. Mineralogy, physics, chemistry, and logical reasoning had combined in their emphasis to engender a veritable vocation of science. The language adopted by its practitioners would have been understood by today's specialists. However, this "renaissance," properly hailed, did little to ameliorate the turmoils of economic, political, and medical life that pervaded the entire human population, whether in the wilds of Florida, New Mexico, or Brazil, or in the capitals of the Commonwealth, Europe, and Asia. Nor did scientists view themselves as agents of cultural interference. There was no link, as yet, between theory and practice, understanding and responsibility, and insight and stewardship. The priestly caste of deep knowledge was, by the very criteria of disinterested objectivity that science had emulated, removed from all these tiring vicissitudes of human civilization.

Yet, Voltaire would speak of that "consolation to the human spirit for the calamities which it will experience in all ages," and he was referring to "philosophy" that he—like Newton whom he greatly admired—considered synonymous with "science" and "scientist." What commiserations? Knowledge itself? Or some other practical domain whereby the fruits of scientific inquiry might trickle down to ease the burden of an existential reality?

By the late nineteenth century, science had begun to embrace its powerful roles, transforming countless discoveries into a policy, a patent, an application, or a profit, where possible. This is to cast no cynical bent to the remarkable history of research, or to knowledge for the sake of knowledge, but to recognize the evolving dependencies between science, government, and industry that arose in the context of complex geopolitical machinations and in the turbulent destinies of over a billion people (the global population at the time of Karl Marx). By the period of relativity, quantum physics, and—most emphatically, World War II—science had become a power unprecedented in the history of human capacity; a new force to rival all others in the puzzle of warring nations and aspiring societies. But whether for Pope Sylvester II, Francis Bacon, or Albert Einstein, the perilous minefield was the same: The fate of knowledge, the fate of the world, and the role of the human heart.

The Present Context and the Twenty-first Century

There was a time, not too long ago, as one of our contributors points out, when the newly described "atom" was more or less a joke in some quarters; a perceived fancy of little likely concern to anybody, ever. Similar scorn would be levied in their day upon electricity, oil, the automobile, even the shopping cart, and—more recently, with echoes still resonating—the personal computer. Even the fuel cell, that ingenious combination of technologies that powered us to the Moon, and back, and now enjoys a status as a likely cornerstone of the alternative energy revolution, witnessed in the aftermath of the Apollo program, a lapse of interest for nearly two crucial decades. Similarly, there were whole centuries when Aristotle was ignored, in particular, his knowledge of biology and intuitive grasp of ecological interdependency. He also recognized the hazards of human overpopulation. His revival in much later centuries coincided with the first stirrings of biodiversity loss; a recognition by some that forests were disappearing, cities becoming polluted, and water fouled. Plato, too, had warned of environmental disruption, citing the destruction of Mediterranean watersheds. Today, Plato and Aristotle are seen to have been philosophers and scientists deeply concerned not merely about Ideal Forms, or Republics, but about real problems in their time. Problems all too with us.

Now, no one takes lightly the revolution in knowledge and technology, the advances and critical importance of all the sciences, and their collective, indeed—urgent—relevancy to the twenty-first century. Hence, the two gatherings of scientists at Anaheim and Budapest—nearly six thousand of them— and the co-attendance by policy experts, corporate leaders, politicians, students, and journalists from all over the world.

The goals of these two august assemblages were many, perhaps overly ambitious as they needed to be. But foremost among their concerns were five consistently articulated points: (1) the need of scientists to listen to one another, and to state their cases clearly and compellingly to the public, to other educators, and to policy makers; (2) the importance that nonscientists pay attention; (3) the inextricable relationship between science and ethics; (4) the ecological crisis that is very real and that must necessarily require the coordinated efforts of nearly all scientists, policy makers, and the public to turn it around in time; and (5) the fact that science has the expertise, the tools, and methodologies— but only if granted the priority, the economic preconditions by governments— to help make this a better world for all.

Each of these imperatives ring loud and clear in the interviews presented in *A Parliament of Science.* Consider Robert Watson describing the struggle to turn the desperate insights of global climate science into workable policies to wake up government and industry with their vested political and economic interests. Or Anthony C. Janetos and Robert May conveying the alarming particulars of a whole new wave—possibly the sixth spasm, so-called—of planetary extinctions, a "peppering of small holocausts" across Earth.

Others, like Rita R. Colwell and Nobel Laureate Leon M. Lederman, speak to the sheer joy of science, and its importance to civilization, as well as to their own personal lives. They each make powerful arguments for an enhanced appreciation of science education, as well as for theoretical research. In the case of Lederman, it was the invisible neutrino that early on absorbed his research; for Colwell, the equally omnipresent and curious bacteria. Other contributors, like Frans B. M. de Waal and Nobel Laureate Joseph Rotblat, have devoted their research to peace. To understanding the mechanisms by which other primates make peace; and to challenging our too easy assumptions and habitual patterns of conflict. Rotblat's voice haunted all those present at the Budapest conference, exerting an unforgettable injunction and putting on notice, in essence, the human race: Make peace, not war; make certain that science is in no way perverted by those who would sooner turn to hatred, division, and killing, than to nurturance, love, and empathy.

In holding science to the highest levels of accountability, Margaret Somerville challenges us to rethink cloning and bioengineering. She writes,

> We have to find some way that we can all personally identify with life; relate to life; and through which we can personally, and as communities, find this sense of deep respect for life that we are prepared to maintain. Out of that comes probably the most important ethical question that we will ask: What should we not do that we now can do?

In other approaches to the debate, M. S. Swaminathan and Ismail Serageldin bring a deep empathy for widespread suffering to the table. How can we minimize pain in this world? The question involves not merely science, and ethics and spirituality, but the most practical considerations for dealing with inequities in global agricultural, health, and human resources. Can cloning be applied in ways that will unanimously serve humanity without violating inalienable human and other animal rights, exacerbating existing social fears and differences, or infringing on the fragile web of biological habitat? And how might traditional wisdom provide additional insights to such questions and dilemmas? Indian ecologist Madhav Gadgil looks at ancient customs and cultural contributions still being made in his native land that suggest additional and often crucial contexts for even framing the debates that engage science.

The debate about global warming is less divisive, more universally acknowledged, but no less troublesome in its implications or confounding resolution. The awesome truth of human overpopulation factors into nearly every prospect of the future, only heightening the confusing array of priorities, and investing each and every one of us with a mission to do our best, in return.

Science, in the end, can not dictate policies of sustainability. Only NGIs— non-governmental individuals, their communities, and elected leaders can collectively do so. It can, according to Ismail Serageldin, feed those eight hundred million who are hungry, and those billions of people without electricity or proper sanitation or clean water, health care, or education. And it must, according to the general chorus of voices throughout *A Parliament of Science* act responsibly to conserve the earth, cherish all life, and pass on a legacy all future generations can live with.

In the end, it is hope in humanity itself around which science must rally. Says Federico Mayor, the Director General of UNESCO at the time of the Anaheim and Budapest conferences, "Human beings are the eyes of the Universe, and these eyes that know what is happening and that design their own future, that is our hope." For Mohammed H. A. Hassan, President of the African Academy of Sciences, this hope is lodged in his own family. He refers to his two daughters, like so many thousands of other young African students, for whom it is his ardent wish that they manage to return home and find the way to utilize their evolving knowledge base "to help foster an African renaissance." For author, professor, and former ambassador Crispin Tickell, his hope is not only grounded in the next generation, but he also looks to the adults of today who "have got to admit," he argues, "that the world in which they were born and in which they are now growing up has got a lot of things wrong with it." Tickell is particularly concerned about humanity's ability to steward and shepherd an interdependent world, one in which life itself has utterly shaped the planet we all must share.

Because so much of science is focused upon life itself, it is not surprising that the scientists who speak out in this volume are all deeply concerned with the future of life on earth. Yechiel Becker, professor of molecular virology at Hebrew University in Jerusalem, considers how science and scientists might intervene to create mechanisms for peace, whether in the Middle East, against bioterrorism, and in regional conflicts everywhere. Similarly, John Durant explains how the public's perceptions of science, and its anxieties about such currents as genetic modification, need to be addressed by scientists who are willing to interact with the public and to make their enterprise accessible and comprehensible. In addition, says Durant, scientists have to take more responsibility for what it is they do, and the power they wield.

This interaction with the public, argues Julia Marton-Lefèvre, is crucial to solving problems. If human behavior needs to change in order to compensate for problems our species alone has inflicted, scientists—who are often in the front trenches of analyzing those problems—must work in partnership with the public.

Ultimately, science will go nowhere if it can not adequately embrace all people. Bruce Alberts, president of the National Academy of Sciences in Washington, D.C., believes that this can only happen when scientists are prepared to share their knowledge with peoples of all nations. To ask questions and seek answers that will be not only of theoretical importance, but useful for people in need, while providing incentives and inspiration to one generation after another of new students who can thrill to the mysteries of the world and find in science multiple avenues all open to them. "Every child [is] a scientist," says Alberts, speaking of new science curricula that provide hands-on experiences; and that give children, in particular, new ways to think, and new and exciting opportunities for becoming effective citizens of the twenty-first century.

I

Biosphere, Ecology, and Sustainability

CRISPIN TICKELL

The Scandal of Unsustainability

Crispin Tickell

Sir Crispin Tickell, GCMG, KCVO is Chancellor of the University of Kent at Canterbury; Trustee of the Natural History Museum and the Royal Botanical Garden Edinburgh; Chairman of the Climate Institute of Washington, D.C.; and President of the Earth Centre in South Yorkshire.

Most of his career was in the British Diplomatic Service. He was Chef de Cabinet to the President of the European Commission (1977–1980), Ambassador to Mexico (1981–1983), Permanent Secretary of the Overseas Development Administration (1984–1987), and British Permanent Representative to the United Nations (1987–1990). He then became Warden of Green College, Oxford (1990–1997), and remains Director of the Green College for Environmental Policy and Understanding. He was President of the Royal Geographical Society (1990–1994), Chairman of the International Institute for Environment and Development (1990–1994), Chairman of the Government's Advisory Committee on the Darwin Initiative (1992–1999), President of the National Society for Clean Air (1997–1999), and Convenor of the Government Panel on Sustainable Development (1994–2000).

He is the author of *Climate Change and World Affairs* (1977, 1986), and *Mary Anning of Lyme Regis* (1996). He has contributed to many books on environmental issues (including human population increase and biodiversity). He is a member of two Government Task Forces: one on Urban Regeneration, and the other on Near Earth Objects. His interests range from business and charities to climate, mountains, pre-Columbian art, and the early history of the Earth.

The way in which we run our societies is endangering the well-being of our planet and its life systems. Future generations will, I believe, look back on our time as a watershed in the relationship between humans and their natural surroundings. Most people have still to recognize an ultimate target of a society with population, resources and environment in the balance. Until they do so, the juggernaut of the conventional wisdom, powered by inertia, vested interests, and unwitting attachment to things as they are, will roll destructively on. If we do not make the change ourselves, nature will do what she has done to over 99 percent of species that have ever lived, and do the job for us.

Sir Crispin, you have spent most of your life tracking the planet's ecological crises. Summing up, where do we stand today, and what are the most glaring problems facing us, speaking as a member of a species with inordinate power over the rest of nature?

I find that of all the subjects that cause people distress about the degradation of the environment, the degradation of biodiversity is the most difficult to get across. People don't really understand what's going on. They think that, yes, the forests are being destroyed, ecosystems are perishing, but I'm all right. I'm getting my food; what's it all about? But in some respects it's probably the most important of all.

Every year the Government Panel on Sustainable Development complains about the depletion of the world's fishing stocks. Every year we are told politely that it's all too difficult or that something is already being done to put it right and we shouldn't be too impatient. In fact, the scandal, and that is what it is, is getting steadily worse. For example, the subsidies put into fisheries is comparable to the price the fish fetch in the markets. At the same time the stocks are depleting worldwide. People don't want to touch this issue because it's complicated. It relates to the Law of the Sea; it refers to the fact that in matters of the sea, we are hunter-gatherers and not cultivators. It raises all sorts of nasty issues that people don't want to confront. Meanwhile the fish stocks of the world are gradually running down and people are being paid to run them down.

> Marine biologists at the UN Food and Agricultural Organization who monitor oceanic fisheries report that nearly all fisheries are now being fished at or beyond capacity. After increasing from 18 million tons in 1950 to nearly 90 million tons in 1990, the oceanic catch has fluctuated around the same level during the level during the last seven years, showing little sign of increase or decrease.
> —Lester R. Brown, *State of the World 1998*, p. 5

The rate of biodiversity loss is probably about a thousand times the natural rate of loss and the replenishment is very, very slow. At least now people are focusing on it. I approach it from the point of someone who doesn't know the right solutions except the need for a much greater respect for life. We need more interdisciplinarity in universities so that people understand the complexities of ecosystems and the vulnerability of the natural world.

Valuing Natural Services

In a world where people tend to calculate things in terms of dollars, biodiversity can be put, however reluctantly, into a slightly commercial framework. A study was conducted at the University of Maryland about two years ago; the goal was to put a value on natural services: all the things that we think we should get for free, like air, water, and soil. It was difficult, but eventually the scientists arrived at an average figure, roughly 33 trillion U.S. dollars per year. They compared that with another very hazy measurement, Gross National Product, and that came out at substantially less, somewhere like 30 trillion U.S. dollars a year. This shows that we're getting for free, and abusing what we get for free, more than what we actually produce. That's a very telling way of looking at it.

> Robert Costanza of the University of Maryland and colleagues from around the world calculated that the current economic value of the world's ecosystems is at least $16–54 trillion per year, exceeding the gross world product of $28 trillion (in 1995 dollars). If every service for each ecosystem type were measured, the figure would be much higher.
> —Janet N. Abramovitz, *State of the World 1998*, p. 37

I think it's wrong to try and put a commercial slant on the protection of bio-diversity, but it does help to bring home what in fact we're dealing with, above all to economists. It shows what the gigantic costs would be if we exhausted resources, destroyed ecosystems, and ruined the resource base on which all our civilizations depend.

The Demographic Winter

If I have to single out one major environmental problem, it is the multiplication of the human species. If the species that was multiplying were crabs or caterpillars or swallows or scorpions, we'd be scared silly. But it's ourselves and it happens relatively slowly, so we don't notice it most of the time. Perhaps a telling way of drawing attention to it is to say that between the Rio Summit in 1992 and the Rio plus Five meeting in New York in 1997, 410 million new people were born. In just five years, we produced more people than lived two thousand years ago at the time of the Roman Empire and the Han Empire in China.

The industrial countries feel relatively secure because their populations are in broad balance. They think that the multiplication of human numbers in other parts of the world is not their concern. They don't take account of the effect it's going to have in generating refugees. When you couple human population increase with environmental degradation, the numbers of refugees are going to multiply at an increasing rate. There will be effects on human health and on the use of resources. Things are going to get worse in the twenty-first century.

Four Ways to Slow Down Population Growth

How can we slow down and then stop this incredible increase of humans? The problems are well illustrated in India. In the North human fertility rates are very high and the population is multiplying. That's in a very poor part of India. Women are particularly disadvantaged. Compare that with the state of Kerala in the South, where women are educated, have full status as citizens, and have a different way of looking at life. That population is broadly in balance.

> Nearly 60 percent of the projected population growth is expected to occur in Asia, which will grow from 3.4 billion people in 1995 to more than 5.4 billion in 2050. By then, China's current population of 1.2 billion is expected to exceed 1.5 billion, while India's is projected to soar from 930 million to 1.53 billion. Over this same time period, the population of the Middle East and North Africa is likely to more than double, while that of sub-Saharan Africa will triple. By 2050, Nigeria alone is expected to have 339 million—more than the entire continent of Africa had 35 years ago.
>
> —Lester R. Brown & Jennifer Mitchell, *State of the World 1998*, p. 174

You can pull back population increase in four ways. The first is by giving women the status they should have and their power of choice. The second one is education, particularly of young women. The third is the affordable availability of contraceptive devices. And the fourth, is to ensure that older generations can look after themselves with proper pension schemes. If you bring those four factors together, I believe that you can gradually slow the rate of human increase and then bring it into balance. Eventually, of course, it will have to decline.

Different countries are choosing different methods. A country like Bangladesh is doing better than people give it credit for. In countries like Pakistan, however, it isn't happening. China introduced the draconian rule of one child per family. That has quite drastic effects on society. You drive through a Chinese village and there aren't nearly so many babies playing in the street. You drive through a Mexican village where the population is still increasing at a giddy rate, and you see children mostly under fifteen.

If you set in motion these measures you are in fact bringing about a whole range of social consequences. You're going to cause the graying of society albeit on a temporary basis. Most governments don't want to think about that because it's obviously very difficult. I suspect that some of them secretly hope that their surplus populations will turn into refugees and go somewhere else. If you couple that with environmental degradation and climate change, you realize you've got a very, very big problem for the next century.

The Urban Crisis and the Sustainability Factor

In Britain I'm a member of a Task Force on Urban Regeneration. It involves looking at what's happened to cities since the Industrial Revolution. I've toured different parts of Britain and seen the decay that has taken place. A rather good analogy is that of an Australian grass known as spinifex. When it seeds, it begins in the middle, the central point, and as it grows, it grows outward. The inner part weakens until it turns into a brown patch. The spinifex expands, using up its resources so the brown patch in the middle becomes more and more like a desert. That's what's happened in the case of many of our cities. They are growing outward and they're invading the countryside.

Take London, for example. It needs 112 times the space of London to sustain London. What the Task Force is doing is to see how a sense of urban identity can be restored, how a human, rather than a megascale, dimension can be restored to human affairs. People will be traveling less to cities in the future; they'll be working more at home. It's not just the poor countries of the world trying to catch up with the rich countries; it's also the industrial world facing the effects of 250 years of industrial revolution and assessing the scars it has inflicted on itself. How do we heal those scars? How are we going to make better use of our resources in the future?

Soon 50 percent of the planet's population will live in cities, and that figure may increase. These tendencies engender positive feedback. If people find cities intolerable, they'll want to move out of them. As the center of the city becomes degraded, many who can afford it want to move outwards. If you live in a big country like the United States or Russia, perhaps it's less of a problem. If you're living on a small island like Britain, then it's a big problem. How are you going to cope, to protect the spaces that are left? How can we resurrect cities so that they are living organisms again?

Each city requires a much larger area of land (carrying capacity) to sustain it than it has in itself. You have to look at cities like superorganisms. They absorb resources, water, timber, stone, oil, and gas, all the rest of it and they emit waste. Rather like a huge anthill. Those cities that are superorganisms are all very vulnerable. If you go on with it too far and you keep on expanding them, they will crash. If you look at the history of human society, there have been roughly thirty civilizations, and all but ours has crashed because of the destruction of their resource base. Nothing is inevitable and we don't have to crash. But we must start shaping things soon if we are to avoid disaster. First people must recognize and understand that there is a problem.

Targets for Development

What does development mean? Is there such a thing as a developing country? Do we simply mean improving standards of living? Then what sort of standards of living? There is a search for producing highly developed mini-Swedens in places where you won't get anything like the resources that sustain the Swedens of this world.

A lot of the problems of so-called development involve striving after what people cannot get. One of my jobs in the past was to be the Permanent Secretary of our Ministry of Overseas Aid. I traveled around the world and my message was that development doesn't mean anything until you make it mean something. If you use the word "change" instead of "development," what paths of change do you want? How are you going to keep your population reasonably fed and doing useful jobs and leading fruitful lives? That's a very difficult question to answer. It doesn't necessarily mean bringing in Western style power stations or football stadiums, or pioneering exhaustion of natural resources. Each country has to choose the way it wants to go.

Development and the Environment

In the case of Kerala, there has certainly been a lot of destruction of the forest and a lot of exhaustion of such resources as fish. The Chinese are realizing after the disastrous floods in 1998 that the reason for the disaster was not so much Mother Nature as the way in which people had exhausted timber resources and damaged top soils. The people of Kerala have to learn that lesson, replant their forests, and achieve a better balance between cultivated land and forestland.

15

In the case of fisheries I must admit a small measure of guilt. I remember that people in one of the Indian states wanted advice on how to improve their fishing takes. We gave them advice only to find that the advice was so readily taken that there were soon no fish. That's not the problem of the big trawlers. It's much more the problems of inshore fishing and the way things get into or out of balance.

> The generation born before midcentury enjoyed a doubling of seafood catch from 8 kilograms per person in 1950 to 17 kilograms in 1990. If the warning advice of marine biologists some 20 years ago had been heeded, and if the world population had been stabilized, consumption would remain at 17 kilograms. Unfortunately, however, while the last generation benefited from a steady growth in the seafood catch per person, the next can expect a steady decline and a rise in seafood prices that will likely last until world population growth will come to a halt.
> —Lester R. Brown, *State of the World 1998*, p. 5

In the same way people complain about the amount of water leakage from pipelines. That's true. But if you cleared up all the leaks, many trees and fields would shrivel up as the leaks themselves play a positive role. There are no easy answers.

Can We Advise an Alternative Path?

A cultural change is required. In a way the industrial West has imposed its standards on the whole world. The world has listened all too readily and now wants to have the same standards itself, but the planet probably can't take it. That doesn't mean that we should say to people: You can't have it. They have to look at their own resources and capabilities and say, What can we achieve ourselves within our own environment? Each part of the world has a different environment. You must include existing industrial countries because they have run into a new range of problems that also render them out of line with their environment.

The Greenhouse Debate

The science of the greenhouse debate is now well established. We have one large country, the United States, where vested interests, such as the Global Climate Coalition, resist that science in order to frustrate change. But the rest of the world is fairly convinced, and the scientific case is broadly accepted.

I've been to meetings with British political leaders and some of the heads of major British companies, and there was no argument about the need to reduce carbon emissions. The scientific arguments have been largely superseded by much more difficult arguments over what you're going to do about the problem.

The Global Climate Coalition brought together an unholy alliance of motor car manufacturers and fossil fuel producers. Two of the biggest oil companies in the world, BP/Amoco and Shell have left that group and they're both putting a lot more money into renewable energy.

Reducing Carbon Emissions

The Intergovernmental Panel on Climate Change that brings together the world's scientists, including many who were highly skeptical, produced its third major assessment at the beginning of the year 2001.

The problem is now less the science but its application. The industrial countries are extremely reluctant to take measures that their scientists tell them are necessary. The quantity of carbon now being put into the atmosphere is liable to cause climate change and may already be doing so. With a population pressure as great as it is and as great as it will be, any form of climate change exposes the vulnerability of human society. Consider, for example, likely shortages of water in the twenty-first century, which the United Nations recently reported upon. It is linked to the greenhouse problem. In Europe, which has been the leader in the climate change debate, nearly every country is taking measures to reduce its carbon emissions.

I think that in Britain we shall meet our target of reducing carbon emissions to below 1990 levels by the year 2010. We've set ourselves another less binding objective, which is to reduce them by twenty percent. Europe as a whole will probably meet its commitments under the Kyoto Protocol (the international agreement to reduce greenhouse emissions, adopted in 1997). The problem is the United States that has less than five percent of the world's population and produces about twenty-five percent of its pollution.

> Children, whose developing lungs are especially vulnerable, are increasingly at risk. A recent examination of 207 cities ranked Mexico City, Beijing, Shanghai, Tehran, and Calcutta as the five worst in terms of exposing children to air pollution. These cities are home to the largest population of young children choking on the worst combination of sulphur dioxide, particulates, and nitrogen oxides. Just by breathing, these children inhale the equivalent of two packs of cigarettes each day.
> —Molly O'Meara, *Vital Signs 1999*, p. 129

So far the United States has tried to evade this issue. I regard most of the so-called mechanisms in the Kyoto Protocol as fundamentally fraudulent. They are designed to give the impression that we are solving the problem while doing extremely little about it. They are anyway highly theoretical. I take a pretty gloomy view of the climate debate.

The Influence of the Third World

Carbon emissions in the United States and Canada are around 5.8 tons per person per year. In Europe this figure varies between 1.8 and 2.6, that's Germany,

Britain, France, and so on. But in China it's just over 1 and in India it's still less than 1. So the great increase in carbon emissions that may come from the South has not yet happened. I believe that the only way in which we can cope with this particular issue is to develop other energy technologies as fast as we can. There are many ways of generating energy that aren't using coal, oil, and gas.

I believe that the Chinese are thinking about building a one-hundred megawatt power station from solar cells. In Britain there is increasing emphasis on wind power, for which Denmark is a demonstration case. Tidal stations are planned and the solar cell industry looks set for take off. The difference between the price of energy generated by these means and generated by conventional means is now narrowing quite substantially.

The fundamental reason why people are going to move to these technologies is that they will see it's in their own self-interest. China has less than five percent of the world's water, and almost 23 percent of the world's population. The work done by the Chinese National Academy of Sciences shows that the risks of climate change for China are very real. There could be less rain, therefore less water and the rain that does come could fall in the wrong places. It could fall on the rather dry soils in the Northwest that are not good at absorbing water.

> Perhaps one of the most underrated issues facing the world as it enters the third millennium is spreading water scarcity. As water use has tripled since midcentury, it has led to massive overpumping. Water tables are falling on every continent—in the southern Great Plains in the southwestern United States, in southern Europe, in North Africa, in the Middle East, in Central Asia, in southern Africa, on the Indian subcontinent and in central and northern China. A matter of growing concern for many governments, water scarcity is often considered separately from food scarcity. But 70 percent of all the water pumped from underground or drawn from rivers is used for irrigation, so we face a future of food scarcity.
>
> —Lester R. Brown, *State of the World 1998*, p. 6

If we say to the Chinese, are you going to continue to be 65 percent dependent on coal, they are likely to say, the sooner we can get away from that the better. What are you doing to help? Part of their answer is nuclear that is a subject in its own right. Part of it is going for other forms of energy.

Countries have got to look at their own self-interest and see what could happen to them if there were to be climate change. It would mean rain in the wrong places, droughts in the wrong places, more extreme events and all the rest of it. I think, for example, that some aid from industrial countries with their scientific base should be directed toward helping other countries understand what could happen if they don't get this right in the next century.

The World Conference on Science

It's a good occasion for interdisciplinarity. It is often a great fault of science that scientists get into their specialized boxes. They don't very often lift the lid and look around. Scientists hate being wrong so they like to be sure of everything. They tend not to venture out into the imprecisions of public debate.

What is the future of science as a component of society? It's very easy for scientists to say that they want more money. It's equally easy for governments to say: no, you've got quite enough already. A lot of the most interesting discoveries have been serendipitous and not the result of dedicated programs. Some scientists spend time chasing ideas that turn out not to be particularly useful or not going to lead anywhere. But they should have the opportunity to do so.

Governments say to themselves: we've got a problem because we are accountable to our taxpayers. Why should we be paying all these people to do things that don't look particularly interesting or that are not even relevant to the problems that look important to us? There is a constant tension between science and government, between science and the funders. I have sympathy with all concerned.

The Unity of Life

The idealism of the young is very important and should be harnessed to this endeavor. But we, the older generation, have got to admit that the world in which they were born and in which they are now growing up has a lot of things wrong with it. We've now got to move into a better direction. That means quite substantial changes in the way we think and the way we treat our natural surroundings. We need a mental revolution by which we think differently.

It's very important to see the unity of life. The most precious thing that we've got is life. The most precious thing that we can understand is the relationship between different living things. What is called the GAIA theory brings together the interdependence of all living things and our sense of responsibility toward other living things. GAIA means looking at the world as a whole, seeing the infinite interconnections, and realizing the role of life in shaping the environment of the planet in which we live.

References

Tickell, Crispin. 1996. *Mary Anning of Lyme Regis.* Lyme Regis: Lyme Regis Philpot Museum.

———. 1986. *Climatic Change and World Affairs.* Lanham, MD: University Press of America.

———. 1978. *Climatic Change and World Affairs.* New York: Pergamon Press.

ROBERT MAY

The Sixth Wave of Extinction

Robert May

Born and trained in Australia as a theoretical physicist/applied mathematician Robert May (Professor Lord May of Oxford) has studied for the past twenty years how populations and communities are structured, and how they respond to change, both natural and human-created. Considered one of the leading mathematical biologists of our time, he is best known among the wider public for his discoveries in the area of chaos theory. He has held professorships at Princeton University and in Sidney and he is a Fellow of the Royal Society and Foreign Member of the U.S. National Academy of Sciences. In 1996 he won the Crafoord Prize, the equivalent of a Nobel Prize for Biology for his population biology work. In 2000 he was appointed to the highly regarded position of President of Britain's Royal Society.

He was Chief Scientific Adviser to the U.K. Government (1995–2000) and Head of the U.K. Office of Science and Technology on leave from a Royal Society Professorship in the Department of Zoology (Oxford University) and at Imperial College (London). Prof. May is and has been, a member of many committees, such as past President of the British Ecological Society, Chairman of the Board of Trustees of the Natural History Museum, and past Trustee of the World Wide Fund (WWF), United Kingdom.

Professor May, there seems to be much debate among those in the biological sciences as to the precise number of species on Earth at this point. And this uncertainty is surely complicated by the ever-increasing numbers of species that are going extinct. Could you shed some light on these questions?

That's the kind of question I would very much steer away from trying to answer because we don't have a clue. I would rather sketch a hierarchy of uncertainties, which brings that statement into sharper focus. How many species have we named and recorded? How many species alive today are currently known to science? No one knows the answer to that simple, factual question to within about ten percent. My best guess would be about one and a half million. It might be as high as 1.7 million and as low as 1.4 million. This is partly made complicated by the fact that without a centralized database we're having trouble resolving synonyms, the same species that's been independently discovered twice. For example, roughly forty percent of the beetles that are known to science, are known from one geographic site, sometimes from one specimen.

The next question is how many species may be alive on Earth today? There is a range of estimates that go from what I think is a rather low estimate of maybe three million, twice the number recorded, to up to tens of millions. The latter estimate I regard as wildly exaggerated and rather unsound, but it is nonetheless arguable. My best guess would be maybe seven million. You could make a very plausible case anywhere in the range of five million to fifteen. Ultimately we don't know within a factor of ten. It's the little unfashionable things, microorganisms in the soil and insects. Arguably the small things that run the world have never been fashionable. If we don't know within a factor of ten how many species are alive on Earth today, anyone who tells you how many are going extinct each day is a fool.

A Sixth Wave of Extinction?

We have to go at it in a different way. Instead of asking about numbers that we don't know, we have to ask of the ones we know reasonably well, birds, mammals, and palm trees; what's the rate at which we've lost species in the last century compared to the average loss rate in the fossil record? Over the last century about a hundred species of birds and mammals have become extinct. Those have certified extinction certificates. It may be more because there may be some in the tropics that we don't know and love enough to have given them a death certificate.

> Of the nearly 10,000 species of birds on the planet, more than 1,000 are officially threatened with extinction. For mammals, where some 1,100 species out of 4,400 are threatened with extinction, the numbers are even more alarming. Among mammals, the 232 species of primates—our closest relatives—are most at risk, with the survival of nearly half of them in question. As our numbers go up, their numbers go down.
>
> —Lester R. Brown, *State of the World 1998,* p. 11

So that extinction rate is about one a year. That doesn't sound like much. There are nine thousand bird species and four or five thousand mammal species. That's only one in ten thousand each year. That means that if you were playing Russian roulette with a one chance in ten thousand of losing each year, it would mean that your life expectancy is roughly 10,000 years. Sounds like a long time, but let's look at the average life expectancy of an animal species in the fossil record over the last 600 million years. It's very variable of course, some are as short as a few hundred thousand years, some tens and tens of millions of years, but on average that's in the range of one million to ten million years.

> The threat to fish may be the greatest of all, with one-third of all species—freshwater and saltwater—now threatened with extinction. In North America, 37 percent of all freshwater species are either threatened or already extinct. In Europe, the figure is 42 percent. In South Africa, two thirds of the 94 fish species are expected to disappear in the absence of special efforts to protect them. In semiarid regions of Mexico, 68 percent of native and endemic species have disappeared. As various life forms disappear, they affect the entire ecosystem and particularly the basic services provided by nature, such as pollination, seed dispersal, insect control, and nutrient cycling. This loss of species is weakening the web of life, and if it continues it could tear huge gaps in its fabric, leading to irreversible changes in the Earth's ecosystem.
>
> —Lester R. Brown, *State of the World 1998,* p. 11

The conclusion is therefore that extinction rates in well-understood groups have speeded up over the last century by a factor of somewhere between a hundred and a thousand, and maybe even more.

As we peer more darkly into the future, we have to use various methods of guessing, but the most popular one is species area curves. There are at least three other ways of making this estimate, which all suggest further speeding up. We now have an estimate that says that over the next century we're going to see species extinction rates accelerate by a factor of ten thousand, give or take a factor of ten. Maybe as little as a thousand, maybe as much as a hundred thousand compared to the average over the sweep of the fossil record. That is a number that puts us clearly on the breaking tip of a sixth great wave of extinction. A wave of extinction that differs from the big five in the fossil record in that it is associated with us, not with natural events.

> Depleting fisheries has ripple effects throughout the marine food chain. In Alaska, for example, pollock catches have nearly tripled since 1986. But since the late 1970s the population of Stellar sea lions, which feed on pollock, has plummeted by 90 percent in western Alaska. In 1990, the National Marine Fisheries Service designated the species as threatened under the Endangered Species Act, and in May 1997 the designation was changed to endangered, an even more serious category. Loss of sea lions has deprived killer whales of their primary source of food. In turn, the whales are now eating sea otters and as a result, sea otter populations have declined by 90 percent since 1990, triggering a surge in their prey, sea urchins.
> —Anne Platt McGinn, *Vital Signs 1999*, p. 36

Extinction in Recent Decades

There has been an understandable tendency in the conservation movement to reach for the most dramatic number which is often the biggest number. I'm a more sober kind of person. I like harder numbers and I'm a little distrustful of some of the more dramatic numbers because I think the reality is dramatic enough without reaching for the biggest number you can give wild credibility to. The number I just gave is an acceleration of a factor of a thousand over background rates during the last century. That's not going to do in thirty percent of species in the next twenty years, but it is dramatic enough to be giving you pause.

Dying Coral Reefs

One of the faults of the estimate I just gave is that it's based on birds, mammals and palm trees and there are worries about extrapolating from one group to another. There will be groups like corals that are particularly sensitive to temperature change. There will be things that are particularly vulnerable and things that are less vulnerable. On the countervailing against that, the things that are particularly vulnerable have experienced much bigger temperature swings over the last ten thousand years, so maybe they'll be more resilient coming back.

I rather just take a sort of soft focus, wide frame shot that says what we're seeing is a sixth great wave of extinction. Quite different from previous events in that it is unambiguously associated with us. Of all plant material that is created each year, we use somewhere between a quarter and a half and that's an event without precedent in the history of the planet. That clearly isn't capable of much greater extinction. In some marine areas perhaps we use as much as thirty percent.

Implications for Policies

I've been talking about one of my own professional interests, but it is not altogether unconnected with my present job. First, I would observe that the British government has been a leader in many of these things. Climate change is higher on many international agendas than biological diversity at the moment. But the British government has both within Europe and more generally been a leader in at least beginning to move that train out of the station. The Kyoto Agreements are a very small first step on a long and difficult road but they are an important first step. Tony Blair's Party Conference address in 1997 was about climate change and its implications for our future. It is a message that is heard clearly in the British government.

John Prescott and Michael Meecher built on the foundation of the previous government. The previous prime minister John Major announced at the Rio Conference on Biodiversity the Darwin Initiative taken by the British government. It is a modest amount of money, but highly leveraged. It is working with developing countries where the biological diversity often is, to help match our expertise to their problems in ways that leave the result in the countries that need it in a nonimperialistic way.

The message I give to the government is one that government has heard and accepts and it is doing its best to deal with. Which is not to say that it's going to cure it tomorrow. But if you looked around the countries of the world there would be few who have a better record on taking first steps to deal with these things than the British.

International Cooperation

I agree entirely with the statement made a few years ago by Edward O. Wilson and Paul R. Ehrlich and the Concerned Scientists suggesting that the problems of biodiversity and habitat loss, pollution, and so on are going to require an unprecedented degree of cooperation among nations in the twenty-first century.

All these issues that involve collaboration between many different nation states are not easy. Yet that's the way the future is going to be. The theme of the Plenary Session of the 1999 AAAS Meeting in Los Angeles was "International Issues in the Twenty-first Century." You can demonstrate that over the last twenty-five years there have been increasing patterns of scientific collaboration among people in every country. That's driven partly because it's a smaller world, but it's also driven by recognition that many of the problems we face are getting too expensive or too geographically encompassing to be dealt with by one country alone.

Whether it's practical problems like climate change, biological diversity, or even fundamental inquiries into the nature of the world, particle physics, and astrophysics. The problems of the twenty-first century are going to have to be dealt with internationally and that is a difficult learning process.

Involving the General Public

In the twenty-first century governments are going to have to do a better job of being seen to make sensible use of science advice and policy making. They are going to have to do that in ways that engage the wider public so that the public has confidence in decisions that are made. I see it as an interlocking set of things that need to happen. The public at large itself is multifaceted, and too often what we call the public is some small set of pressure groups. If people at large feel they are heard and if people at large can see that the government is hearing all sides of a case letting ideas contend, and if government provides leadership and direction that emerge from that, then I do have hope.

Hope for the Future

The Climate Change Convention in Kyoto is a small first step toward addressing the problem. Taking the first step is probably the most difficult thing to do and I read hope into that beginning. Some would call me naive perhaps. I read hope into the accomplishment of actually making a beginning. Getting countries to begin tentatively to agree to deal with the problem in a collegiate fashion.

It's going to take a long time. Some of the nonbinding aspects can be overestimated. For example, the United States still isn't signed up to the Biodiversity Convention. At the same time I think it's fair to observe that meanwhile they have done a great deal to promote efforts in that direction.

ANTHONY C. JANETOS

Protecting the Biosphere

Anthony C. Janetos

Dr. Anthony C. Janetos is Senior Vice President and Chief of Program at the World Resources Institute, an independent policy research institute that seeks practical solutions to problems of environment and sustainable development. Dr. Janetos was trained in ecology at Harvard and Princeton Universities, and previously managed scientific research programs at the U.S. National Aeronautics and Space Agency (NASA). He has been actively involved in both national and international scientific assessment processes that contribute sound scientific information on difficult issues of environmental policy. He is currently a co-chair of the U.S. National Assessment of the Potential Consequences of Climate Variability and Change, and a leading author of the Intergovernmental Panel on Climate Change (IPCC) Special Report on Land-Use Change and Forestry.

Dr. Janetos, you have spent most of your career looking at sustainable development from a global perspective. How did you come to embrace this professional path? What motivated you?

I started my scientific career as an ecologist with an interest in mathematical models, and found that intellectually extremely stimulating. I did both theoretical work and field experiments and I suppose I could have happily gone on for a long time in that way. But early in my career I got interested in the interfaces that science has with environmental policy and decision making. As a result I have been fortunate to work on such issues as air pollution, forest health and deforestation, and climate change and biodiversity.

The Global Loss of Biodiversity

Two of the defining environmental issues of our time are the loss of biological diversity and human-driven changes in the climate system. All scientific evidence indicates that we are looking at changes in the biological fabric of the planet that are perhaps in human experience. There really is no debate about what's driving them. We are driving them, our societies are driving them, by the way we use land, and by the way we use resources. Are we going to leave a world that maintains the richness of its biological heritage for our children and their children, that they can manage and derive wealth from it in the same way that our societies have been able to do?

> The fastest growth in greenhouse gas emissions in recent years has been in the developing world, where industrialization is still gathering speed. By 1996, carbon emissions in developing countries were 44 percent over 1990 levels, and 71 percent over 1986 levels. Rapid economic growth, particularly in East Asia and Latin America, is driving emissions up as growing numbers of people are able to afford home appliances, motorcycles, cars, and other energy-intensive amenities of a "modern"

lifestyle. The International Energy Agency projects that without additional policy initiatives, global carbon emissions from fossil fuels will exceed 1990 levels by 17 percent in 2000 and by 40 percent in 2010, reaching 9 billion tons per year.
—Christopher Flavin & Seth Dunn, *State of the World 1998*, p. 115

We have a very imperfect understanding of the potential consequences of the observed changes. We know these changes are large and that they are happening. We think that many of the potential consequences are nearly or in some cases completely irreversible. I think it is enormously important to understand the quantitative nature of the decisions that are being made to address the loss of this biological richness.

Biological Richness and Resilience

There has been a lot of scientific work suggesting that the biological richness of ecosystems confers on them a certain resilience to other environmental stresses. That richness can be eroded, either because we simplify ecosystems or because there have been too many invasive species. So one of the major concerns is that if these changes continue, ecosystems' ability to resist a whole suite of environmental stresses is reduced. We could be faced with an enormous simplification of systems and interrupt the flow of goods and ecological services that our societies depend on.

North-South Differences

One of the challenges recognized in the Biodiversity Convention (adopted at the 1992 Earth Summit in Rio de Janeiro; it is an agreement on maintaining the world's biodiversity), is how countries are able to appropriate the value of their biological resources. How can they be valued? What are the mechanisms for recognizing that value, and ensuring that there is equitable sharing of it?

Many of the pressures on biological resources are the result of the demand of industrial societies for commodities: fiber, wood, and so on. One would like to be able to meet that demand in a way that does not exacerbate existing problems. By itself, this would not solve the underlying problems. As important is finding ways to meet the appropriately growing demands of the South in equitable ways. Therefore, policy solutions have to be global, while taking national circumstances and needs into account.

Recognizing the limits of natural systems is often seen as a call for no growth, but the issue is not growth versus no growth. The question is, What kind of growth? And where? Growth based on renewable energy may be able to continue for some time, while that based on fossil fuels is ultimately limited by remaining reserves, but more immediately by potentially unacceptable climate disruption. Similarly, a reuse/recycle economy can grow much larger than a throwaway economy without imposing excessive demands on Earth's ecosystem. Growth in the information economy puts minimal pressure on the Earth's natural resources, especially compared with heavy industry, a common source of past growth. Within agriculture, huge growth is needed to satisfy future food needs in developing countries but not in industrial ones, where population has stabilized and where diets are already sated with live-stock products.

—Lester Brown & Jennifer Mitchell, *State of the World 1998*, p. 170

The World Resources Institute

The World Resources Institute (WRI) is an international policy research institute based in the United States. We try to be creative in looking at the practical implementation of important policy issues. We take a consensus-building approach, a catalytic approach to get the views of multiple stakeholders into important policy discussions and try to find practical ways to move forward.

One of our particular challenges is to find ways to use scientific information to inform policy debates. A number of our programs look at the use of satellite information or geographic information systems, and how those kinds of technologies can help policy communities visualize problems and therefore hopefully visualize solutions. Almost everybody knows how to read a map. Being able to visualize a problem such as the fate of forested ecosystems, their extent and how they've changed over time is a very powerful incentive to spur debate and discussion on the effects of different future policies.

We work all over the world and are supported by a variety of international sources. One of the challenges that we face is how to bring problems and issues in the developing world into global policy debates in a constructive way.

When WRI began fifteen years ago, there were very few institutions that looked at global environmental issues. There are many more now. This is a sign of success, that the global issues are important, that governments, private corporations, and civil society are all taking them seriously as issues to be resolved.

Our Major Challenge

I would not presume to say that one environmental or developmental issue is more important than any other. The challenge of managing the earth's biological richness, preventing the potentially dangerous interference in the climate system, while sustaining, and even enhancing human livelihood, may be one of the biggest overall challenges for the next generation. Doing that in the context of enhancing human livelihood I think is one of the keys. We can't pretend to have solutions that don't do that; they won't last.

> The good news is that we know what an environmentally sustainable economy would look like. We have the technologies needed to build such an economy. And we know that the key to getting from here to there lies in restructuring the tax system, decreasing personal and corporate income taxes while increasing taxes on environmentally destructive activities. The challenge is to convince enough people of the need to do this in order to make it happen.
> —Lester R. Brown & Jennifer Mitchell, *State of the World 1998*, p. 169

Our children are going to live in a world with a lot more people, perhaps as many as 50 percent more. How to provide basic human needs such as drinkable water and food security, and how to couple these with environmental protection, sustainable development, and increasing prosperity are major challenges. I believe that it is possible to enhance people's lives while protecting the biological wealth and the natural resource base on which all of our societies ultimately depend. That is really one of the central challenges for the coming generations.

What Are the Prospects?

If you go into this challenge thinking that solutions are simply going to be too difficult or too onerous for people to accept, that's not productive in the end. You have to be realistic about how hard the solutions are, but you also have to be optimistic. It's going to be a learning process.

If you look at the kinds of social and environmental changes that we've seen even over the last thirty years, I think that one has to be optimistic that change is possible. If we're clever and persistent we can begin to effect some of these changes.

> One way to boost the energy efficiency of the global economy is to shift from a throw-away economy to a reuse/recycle economy. Another obvious area that needs improvement is the energy-intensive automobile-centered transportation systems of industrial societies, which are extraordinarily inefficient not only in energy us, but in the congestion they produce, which leads to an inefficient use of labor as well. In congested London today, average automobile speed is similar to that of the horse-drawn carriage of a century ago. In Bangkok, the typical motorist now spends the equivalent of 44 working days a year sitting in traffic jams.
> —Lester R. Brown & Jennifer Mitchell, *State of the World 1998*, p. 178

The Challenge for Scientists

One of the main challenges for scientists is how they can be effective in communicating what they know both to society at large and to policy makers.

That's a tremendously difficult challenge. Scientists are not trained to do that. Yet the kind of investment that societies have made in scientists and in science overall actually puts us in a position of special responsibility. There's a responsibility to give back and to play a role that is appropriate but is also helpful in moving these issues forward.

My strong personal belief is that science needs to move from a model of only being driven by curiosity to a model of being driven both by curiosity and a sense of societal responsibility. It's a hard transition, and one in which the environment is a critical area for the twenty-first century.

The World Conference on Science for the Twenty-first Century

One of the interesting things about this meeting is its global nature. The conveners have been very careful to get a diversity of views. Another interesting aspect of this meeting is the attempt to get young scientists involved from around the world. They have not necessarily had as much experience, but still I think they have something really important to offer. Their aspirations for scientific careers differ from what other generations of scientists have experienced. That give-and-take has been really quite interesting.

My hope is that there will be an agenda for the appropriate participation of scientists in the decision-making processes that will actually be acted on. Many conferences write declarations, wonderful statements of consensus and opinion. I suppose that there's a library somewhere that has a whole shelf of such declarations. But one of the best outcomes would be if we had a statement of consensus that could actually be acted on and could result in tangible activities by the scientific community to assist with this kind of complex decision making.

Not only are we not taxing environmentally destructive activities, [but] some of these efforts are actually being subsidized. More than $600 billion a year of taxpayers' money is spent by governments to subsidize deforestation, overfishing, the burning of fossil fuels, the use of virgin raw materials, and other environmentally destructive activities. Governmental subsidies for fishing have boosted the capacity of the world's fishing fleet to twice the sustainable yield of oceanic fisheries.

Says the Earth Council "There is something unbelievable about the world spending hundreds of billions of dollars annually to subsidize its own destruction."
—Lester R. Brown & Jennifer Mitchell, *State of the World 1998*, p. 182

ROBERT WATSON

Where Science and Policy Meet:
The Search for a Common Language

Robert Watson

Dr. Robert Watson is currently the World Bank's Chief Scientist and Director for Environmentally and Socially Sustainable Development. Dr. Watson joined the World Bank as its Chief Environmental Scientist in 1996. Before that, he was Associate Director for Environment in the Office of Science and Technology Policy in the Executive Office of the President at the White House, and Director of the Science Division and Chief Scientist for the Office of the Mission to Planet Earth at the National Aeronautics and Space Agency (NASA). Dr. Watson was also the first Chair of the Scientific and Technical Panel of the Global Environment Facility and a key negotiator for global environment conventions.

Dr. Watson, there has been a fast learning curve with respect to the perils of the Greenhouse Effect. But the debate over solutions is apparently far from over. Where do you stand on the issues, and what do you suggest we as a species do about the problems?

The scientific evidence suggests that humans are changing the Earth's climate, although it must be recognized that there are uncertainties in our scientific knowledge. The key question is not whether climate will change, which is almost certainly the case, but rather by how much, when, and where? Another question is, how much will the projected changes in climate affect issues that society cares about most?

I chair the United Nations Intergovernmental Panel on Climate Change, which assesses how human activities are changing the composition of the Earth's atmosphere and whether these changes are affecting the Earth's climate system. It also assesses the implications of climate change on human health, water resources, agricultural production, ecological systems, and sea level rise. Lastly it assesses the options for society to mitigate the projected increases in greenhouse gases by changing energy technologies, energy policies and land use patterns, forestry and agriculture.

> The mainstream scientific community, as represented by the 2,500 scientists on the Intergovernmental Panel for Climate Change, is quite clear on the need to reduce carbon emissions. The Alliance of Small Island States, a group of some 36 islands countries that feel particularly vulnerable to rising sea levels and more powerful storms, is also actively pressing for a reduction in global carbon emissions.
> —Lester R. Brown & Jennifer Mitchell, *State of the World 1998*, p. 184

The magnitude of climate change is projected to vary across the globe. The land areas are projected to warm more than the oceans, and the very high latitudes in the Northern hemisphere are projected to warm more than the tropics. The two continents that are projected to warm most significantly with adverse effects for society are Africa and Latin America. The temperatures are projected to increase and the amount of precipitation in the northern and southern parts of Africa is projected to decrease. This would lead to reductions in the availability of fresh water, especially in the arid and semi-arid regions; reductions in agricultural productivity; changes in the structure and function of terrestrial and marine ecological systems; and increases in the incidence of vector-borne diseases such as dengue fever and malaria. Sea level would increase significantly in certain areas, especially impacting the low lying deltaic areas such as in Bangladesh, Egypt, China, and in the very small low-lying island states such as the Marshall and Maldives islands.

Can We Stop Global Warming?

I believe that the energy services needed for development and poverty alleviation can be obtained in a much cleaner way, by changing energy technologies and policies. Many countries in the world subsidize fossil fuels, which inhibits the clean production and efficient use of energy. Few countries incorporate the social costs of environmental degradation into fuel prices, which would provide a more level playing field for environmentally clean technologies. Natural gas can be used instead of coal to generate electricity, and greater use can be made of modern renewable energy technologies such as solar energy, wind power, and modern biomass. There are many technological changes that could be used to produce energy in a much cleaner way. Not only would it help to protect the Earth's climate system, but it would also improve air quality in most cities and reduce the levels of acid deposition, which adversely affects ecological systems. Many cities in Asia are already "choking" to death because of particulates in the air, and air quality in most large cities in developing countries is projected to worsen in the decades ahead. The problems of global climate change and indoor and outdoor air pollution can, in many instances, be addressed simultaneously.

My Own Priorities

Addressing local air quality issues and global climate change would require policy reform and the development of new cost-effective environmentally friendly technologies. For example, subsidies for fossil fuels would have to be decreased and eventually eliminated to allow renewable energies to penetrate the market and to encourage the efficient use of energy. Developed countries should substantially increase public and private sector funding for research and development into improved energy technologies such that energy production and use could be environmentally more friendly and realized at a much lower cost.

Climate change is affecting the world's biological diversity. The geographic boundaries of ecological systems are shifting and the richness of the flora and fauna in forested systems and coral reefs is decreasing. We must recognize and improve our understanding of the linkages between climate change and other important environmental issues such as loss of biological diversity, land degradation in arid and semi-arid areas, and water and forest resources.

In addition, we need to understand the interlinkages among the local (air and water quality), regional (acid deposition) and global environmental issues, and how policies can be designed and technologies utilized that can simultaneously address climate change, loss of biodiversity, and land degradation while trying to meet basic human needs such as adequate food, clean water, energy, and a healthy environment.

> While it is not particularly surprising that China's total consumption of some basic resources has now overtaken that of the United States, given its population size, it is startling that it has surpassed the United States in consumption per person of some basic goods such as pork and eggs. Although China's grain use per person, both direct and indirect, is still only some 300 kilograms compared with roughly 800 kilograms in the United States, this is up from 200 kilograms in 1978. As a result, consumption of all grain in China now totals 380 million tons, compared with 245 million tons in the United States. As incomes continue to rise in China, so too will grain consumption per person.
>
> —Lester R. Brown, *State of the World 1998*, p. 12

Developing Countries Are Hit the Hardest

While all parts of the world are projected to experience changes in climate, the most severe impacts of climate change are expected to occur in developing countries. In particular, sea level rise is especially important for small island states and low-lying deltaic areas; and the continent of Africa will be the most vulnerable because of decreases in fresh water and agricultural production and increases in vector borne diseases such as dengue fever and malaria.

In most cases developing countries tend to be more vulnerable than developed countries as they are poorer, more dependent upon natural resources, are technologically less sophisticated, and have less infrastructure and institutions to adapt to climate change. So while most of the greenhouse gases have been emitted from the industrialized world, the countries most affected by climate change will be the developing countries. And, unfortunately, probably the poorest of the poor people in those countries will be most affected.

The Potential for North-South Tensions

I'm an optimist in that I believe that developed and the developing countries can work together on issues like climate change. It is quite clear that the developed countries cannot, in the short term, ask developing countries to assume additional costs to avoid climate change given that developing countries urgently need more cheap energy in order to develop and because developed countries have been the major emitters of greenhouse gases to date. Indeed, the Convention for Climate Change recognizes that all parties have obligations, but they are differentiated, depending upon their special circumstances. However, developing countries can reform their energy policies and use the very best technologies available and leapfrog the technologies that have been used in countries belonging to the Organization for Economic Cooperation and Development (OECD) and in other parts of the developed world, which have led to outdoor air pollution and emissions of greenhouse gases. I believe that the different perspectives of developed and developing countries can be resolved before climate change becomes an even more serious problem.

Predicting the Earth's Climate Change

Temperatures and sea level are projected to rise, and the spatial and temporal patterns of precipitation are expected to change. These are the three key changes in the Earth's climate system that need to be considered. The further out in time, the more the Earth's climate is projected to change, but also the more uncertain our projections are.

Models typically project how the Earth's climate could change during the next one hundred years. These projections are, *interalia,* dependent upon our assumptions of how the world's population will grow; how the world's economy will develop; what energy technologies are likely to emerge; the cost of energy; and how energy will be used.

One key question is whether we are already seeing changes in the Earth's climate. The answer is yes: over the last century temperatures have certainly increased throughout the world, precipitation patterns have changed, and we're tending to see more floods and more droughts. The question is whether these changes are due to human activities. While we cannot prove beyond doubt that the long-term changes in temperature and precipitation, the recent floods in Eastern Europe, Bangladesh, and China, and the very dry weather that led to forest fires in Indonesia and Brazil are due to human-induced climate change, the observations are consistent with the "global warming" theory. Again I should stress there are uncertainties. The challenge for the policy makers is to decide how to address issues such as climate change while recognizing that scientific uncertainty exists.

> The rapid warming of the last 25 years is greater than that of any other period since the beginning of instrumental temperature measurements. NASA scientists believe that the accelerated buildup of greenhouse gases in the atmosphere is the best explanation for the warming.

Scientists reported accelerated melting of glaciers in many parts of the world last year—in the Arctic and Antarctic, as well as in many mountainous regions such as the Andes and the Qinghai-Tibet plateau. A new study published in early 1999 indicates that the huge Greenland ice sheet has been shrinking rapidly, losing nearly 1 meter (three feet) just since 1993 in some areas.

—Christopher Flavin, *Vital Signs 1999*, p. 58

New Energy Technologies

The challenge is to develop carbon-free technologies that produce energy as cheap as fossil fuel technologies. The most promising tend to be solar energy, wind energy, and modern biomass, where fast growing trees or grasses are used instead of fossil fuels to produce energy.

More efficient use of energy is needed, for example, more efficient transportation, better land use planning, and increased use of mass transportation. If the whole world develops in the way that Europe and the United States have developed, where every family has at least one if not two cars, we will see cities absolutely blocked with cars, citizens choking on the exhaust of cars, and a significant increase in the emissions of greenhouse gases. The technological challenge is more efficient cars, houses, industrial processes, and thinking more strategically about land use patterns and the other ways in which society lives.

Developing countries hold a 41 percent share of global carbon emissions, and saw a 39.1 percent rise in output between 1990 and 1998. But the convergence of industrial and developing-country shares masks major disparities in historical and per capita rates. Industrial and former Eastern Bloc nations remain responsible for 75 percent of the carbon emitted into the atmosphere since 1950. The average emissions of 1 American equal that of 7 Chinese, 24 Nigerians, 31 Pakistanis or hundreds of Somalis. The richest fifth of the world accounts for 63 percent of emissions; the poorest fifth countries just 2 percent.

—Seth Dunn, *Vital Signs 1999*, p. 60

The Role of International Bodies

International policy and scientific bodies are essential to address global environmental issues such as human-induced climate change. The United Nations Framework Convention for Climate Change (UNFCCC), which was signed at the Earth Summit in Brazil in 1992, has now been ratified by over one hundred fifty countries. The ultimate goal of the UNFCCC is the stabilization of the atmospheric concentrations of greenhouse gas concentrations at a level that will avoid dangerous human-induced perturbation to the Earth's climate system. Associated with this framework convention is the Kyoto Protocol, which was negotiated in 1997 and mandates developed countries and countries with economies in transition to reduce their greenhouse gas emissions, on average, by 5.2 percent below their 1990 level during the first commitment period, 2008–2012.

In 1988 the United Nations established the Intergovernmental Panel on Climate Change under the auspices of the World Meteorological Organization and the United Nations Environment Program. The IPCC assesses the current scientific, technical, and economic understanding of all aspects of climate change, thus providing governments with a consensus perspective of what is known and unknown. IPCC assessments are prepared and reviewed by experts from over one hundred countries from universities, governments, industry, and environmental organizations. The combination of the UNFCCC and the IPCC provide the international institutions to move the climate change issue forward.

International Assessment Processes

International policy formulation on complex environmental issues involving the global commons such as stratospheric ozone depletion, human-induced climate change, and persistent organic pollutants, or issues of global concern such as loss of biological diversity, land degradation and desertification, water resources, and sustainable forestry, require independent international assessments of the current state of knowledge. These assessments, which complement national assessments, need to involve experts from all countries and from all relevant disciplines (natural scientists, social scientists, and technologists) and a wide range of stakeholders (universities, governments, and the private sector). Peer review is an integral and essential component of the preparation of these assessments. One of the key challenges is to understand the needs of the policymakers to ensure that the assessments are demand-driven not supply-driven, and are policy relevant, but not policy prescriptive. The assessments must be scientifically rigorous, with the key conclusions and uncertainties understandable to the policymakers.

The international WMO/UNEP assessments of stratospheric ozone depletion provided the scientific, technical, and economic underpinning for the Vienna Convention on Stratospheric Ozone and its associated Montreal Protocol on Substances that Deplete the Ozone Layer, including the subsequent amendments and adjustments (London, Copenhagen, and Vienna). Scientific information demonstrated that the release of long-lived chlorine- and bromine-containing chemicals into the atmosphere was destroying the Earth's fragile stratospheric ozone layer with adverse effects on human health. The technical and economic assessments demonstrated that cost-effective solutions were obtainable.

Shaken by an increase in weather-related insurance claims from $17 billion during the 1980s to $66 billion thus far during the 1990s, the insurance industry is

urging a reduction in carbon emissions—in effect, a reduction in fossil fuels. Some 60 of the world's leading insurance companies have signed a statement urging governments to move in this direction, marking perhaps the first time in history that one major industry has pressured governments to reduce the output of another major industry.
—Lester R. Brown & Jennifer Mitchell, *State of the World 1998,* p. 184–185

The IPCC is providing comparable scientific, technical, and economic assessments of the current state of knowledge on the more scientifically, technically, economically, and politically complex issue of climate change.

Similar international assessment processes could facilitate progress on a number of other critical environmental and ethical issues, such as cloning, biologically modified organisms, disrupters of the hormone system, and persistent organic pollutants. In addition, an international assessment of ecosystem condition, past, present, and future, could simultaneously provide invaluable information for the Convention on Biological Diversity, the Convention to Combat Desertification and Ramsar (the wetlands convention).

A key challenge is to enhance the scientific and technical capacity in developing countries in order that they can play a full role in these international assessments and develop and implement national sustainable development plans.

The Dialogue between Scientists and Policy Makers

Scientists have to learn to explain the key uncertainties in a manner that policy makers find most useful, i.e., we must learn to explain the policy implications of scientific, technical, and economic uncertainties. Too often we fail to explain the implications of the uncertainties, erroneously believing that policymakers do not know how to handle uncertainties. Yet policy makers and politicians make decisions of monumental importance every day with incredible uncertainty. For example, the average politician is used to dealing with uncertainty and risk in defense or in economic policies.

For example, many governments spend a significant amount of money and percentage of their Gross National Product (GNP) on defense, e.g., annual defense expenditures in the United States are about $300 billion (1999). It's not that the U.S. government expects to be invaded tomorrow, but given the importance of national defense they are willing to invest significant amounts of money, whether they think there will be a major catastrophe or not. Politicians also make many decisions based on economic projections that are often proved wrong by the time the policies are implemented.

However, when policymakers address environmental issues, they appear to expect perfect knowledge. In other words, they appear to believe that science is absolute. Unfortunately our knowledge regarding these complex global environmental issues is no more absolute than for many of the other issues they deal with. So while policy makers can deal with uncertainty, they have to be able to frame their policy question to us in a more precise manner and we have to learn how to communicate the implications of knowledge and uncertainty to them.

Climate Change Is a Global Concern

Every country, to a different degree, will have to address the issue of climate change by limiting their emissions of greenhouse gases over the next fifty or so years. This will require changes in energy and land-use technologies, practices, and policies. To realize significant reductions in greenhouse gases over the next fifty years will require immediate action. In addition, countries will have to adapt to projected changes in climate, which is expected to affect water resources, agricultural production, forests, fisheries, natural ecological systems, human health, and human settlements. Even though developing countries are the most vulnerable, certain sectors in developed countries are also vulnerable.

In the 1995 IPCC Second Assessment Report we projected that many tropical countries would become drier, agricultural production in the tropics would decrease, two-thirds of the tree species in the Boreal forest systems across the Arctic and Sub-Arctic areas would no longer be viable, and that sea level rise will affect low lying deltaic areas and small island states.

Many parts of the United States, especially the South West, are quite arid and semi-arid and climate change is likely to exacerbate this condition. Even though the United States has the financial infrastructure to deal with a lack of water resources in the South West, climate change could make water resource management even more of a challenge. While climate change is projected to adversely affect many sectors in most countries, there may be some countries where climate change is likely to have a positive effect. For example, warmer summer temperatures will result in a longer growing season for agriculture in the mid- and high latitudes, therefore agricultural production will probably increase, for example in the former Soviet Union and in Canada.

The developed world has recognized that climate change is a serious global issue, and they have to take the first step to reducing greenhouse gas emissions. However, in the longer term, all countries will have to revise their energy and land use policies and technologies.

Looking Back at Rio 1992

Some people viewed Rio as a complete success, others a complete failure. I look at Rio as very a important event that stimulated the thinking of governments, industry, and civil society about the issue of environment and development. It raised the awareness that if we carry on with our current development pathways, we are likely to have severe environmental degradation and economic development will not be sustainable. It was also recognized that with the right policies and technologies, there should be no dichotomy between economic growth and environmental protection.

However, many of the grandiose schemes developed in Rio have not come to fruition. There was a recognition that it was important to improve the science and technology infrastructure in developing countries for both economic growth and environmental protection. Unfortunately, if anything, there is less money being spent on science and technology in developing countries today than in 1992. In addition, there has also been a decline in funding for scientific research in some developed countries. For example, with the exception of Japan, there has been a significant decline in both public and private funding for energy research.

Lastly, many excellent national sustainable development action plans were developed, but few, if any, were fully implemented. Again, the awareness is high, but in many respects the rhetoric is ahead of action.

Science for the Twenty-first Century

The *World Conference on Science for the Twenty-first Century* provides an excellent opportunity to demonstrate the contribution of science and technology to a better world in the twenty-first century. It can also show how natural and social scientists from OECD countries, developing countries, and countries with economies in transition can work together to provide the scientific and technical underpinning for a more sustainable world.

The outcomes of the Conference, the *Declaration on Science* and the *Framework for Action* are important but the commitment by governments and the private sector to enhance funding for science and technology in both developed and developing countries over the next two, five, and ten years is going to be critical. Hopefully, the delegates will go home with new contacts and a renewed enthusiasm about the role of science and technology in the development process. I also hope that the scientific community can learn from the politicians and from each other about what's important in realizing a more sustainable and equitable world. We need to address a number of difficult issues, for example, are all fields of science and technology equally important or are some more likely to be more important to move the development process forward? Hopefully there will be an enhanced dialogue between the scientific community, the private sector, and with government representatives. How we all think about these issues over the coming years is probably going to be most important. Science and technology is likely to play an even more important role in the coming century than it did in the last century; hence it is critical that governments commit to sustained funding for science and technology in both developed and developing countries.

MADHAV GADGIL

Ecology:
Modern Science and Traditional Wisdom

Madhav Gadgil

Dr. Madhav Gadgil is one of the most distinguished population biologists in the world and a leader in Indian science. Dr. Gadgil studied biology at the Universities of Poona and Bombay before doing a Ph.D. thesis in the area of Mathematical Ecology at Harvard University. He has been an IBM Fellow of the Computing Centre, Research Fellow in Applied Mathematics, and a Lecturer in Biology at Harvard University.

Since 1973 he has worked at the Indian Institute of Science at Bangalore. He has also served as a Visiting Professor at Stanford University and as a Distinguished Visiting Lecturer at the University of California at Berkeley. His research interests include population biology, conservation biology, human ecology, and ecological history and he has published over one hundred fifty research papers and two books. Two of his research papers have been recognized as citation classics. He has cowritten (with Ramachandra Guha) *This Fissured Land: An Ecological History of India* (1992), and *Ecology and Equity: The Use and Abuse of Nature in Contemporary India* (1995).

He is currently Professor at the Centre for Ecological Sciences, Indian Institute of Science, Bangalore. In addition to that position, he is Chair of the Biodiversity Unit at the Jawaharlal Nehru Centre for Advanced Scientific Research and Chair of the Scientific and Technical Advisory Panel of the Global Environment Facility. He was elected Foreign Associate of the U.S. National Academy of Sciences in 1993.

Professor Gadgil, there has long been talk of an economic divide between North and South. You have extended that metaphor to include the myriad ecological implications of that growing gulf between the wealthy and the disenfranchised populations of the world. Could you expound upon those issues?

In my book *Ecology and Equity* I have analyzed the social context of ecological problems. There is a major division in society in terms of interests of people. In societies like India and many other developing countries there is a polarization of the population into two kinds of people. There are people who have been called ecosystem people who depend to a very substantial degree on resources from their own ecosystems. The other group are people who are dependent on markets, who have resources brought to them from all over the world. They have been called biosphere people, because they are using the entire biosphere instead of the restricted local ecosystems for meeting their requirements.

When the environment degrades the effect is very different for these two groups of people. The ecosystem people are obviously deeply affected, while the biosphere people can always get resources from another part of India or maybe from elsewhere in the world. So these groups have very different attitudes and approaches toward environmental issues and this to me is a central problem.

The biosphere people are in the decision making position so they can always pass on the environmental costs of their actions to some group or other of ecosystem people. In consequence the strategies for resource use do not have the concurrence or the cooperation of the local ecosystem people. The result is a very undisciplined and highly non-sustainable use of resources.

We must bring on board the ecosystem people who have a genuine stake in the health of the local environment and who also have a deep detailed understanding of the ecological situation in their own localities.

Environmental Problems in India

The most basic issue is water: the overuse of groundwater, the pollution of surface waters, both fresh waters and the sea, and the consequent destruction of fisheries. In my fieldwork I have noticed that water is what concerns most of the local people. It is also becoming a problem for the urban people, because there are not quite adequate quantities of water supply to the cities due to poor management of water resources. Finally there are issues relating to land degradation, devegetation, and deforestation.

There has been quite a bit of environmental action over the last twenty years. Most of it is still centralized. There is so much locality-specific variation and time-specific variation that unless that is taken into account you have serious difficulties. Also the local people should be brought on board in decision making.

What Kind of Science Is Ecology?

Ecologists try to develop a quantitative predictive science. However, I have been struck over the years with the fact that the ability of ecologists to come up with scientific insights that have real implications for ecological management at the ground level remains very limited. This is because these are highly complex systems where our ability to experiment or to replicate is very limited. We cannot apply the traditional methodology of science that holds for simple physical or chemical systems.

What then becomes very important is the detailed local level of understanding of the ecological histories. Since you cannot undertake the kind of experiments which traditionally physicists or chemists can do, we have to rely on natural experiments. The local ecosystem people are familiar with the whole range of experiments that take place naturally. I have found that over the generations they have come up with very impressive levels of understanding of the local ecological situation.

Elephants and Ficus Trees

I have been working for many years in the South Indian Hills where they have wild and tame elephants. They are all captured from the wilderness and are now being maintained in a variety of wildlife sanctuaries, for tourism and for timber operations. About twenty years ago, I heard the elephant men talk about the fact that elephants in the wildlife sanctuaries were being fed on leaves of Ficus trees. Ficus or fig trees are an important genus of trees in India, widely protected not only in India, but also in Southeast Asia and Africa.

These people said that Ficus produces fruit in certain months of the year when no other trees are in fruit. Therefore, a Ficus is an important resource for monkeys and squirrels and lopping leaves of Ficus is damaging in a wildlife sanctuary.

A few years later John Terborgh, a very distinguished American ecologist, wrote a paper based on his study in Manaus, Peru, making exactly the same point and he coined the expression *keystone resources.* Ficus trees are keystone resources because they support fruit eating birds and other mammals in seasons when no other fruit is available. Terborgh's paper has become very famous. Our elephant men were talking about exactly the same effect many years before that paper was published and this idea gained currency in ecology.

The Challenge of Folk Science

Local people have many interesting notions of how the ecological system works. A very important challenge before science is to develop a new synthesis between this folk science and formal science. This must be done if we are to properly manage our complex ecosystems. Local people are motivated to properly manage their own environment and there are very successful experiments in India like the Water Users Associations and the village forest committees.

There is also the issue of understanding ecological processes within the local communities. Most of the so-called ethno-biological work has focused very narrowly on documenting medicinal uses of plants and then perhaps pharmaceutical industries appropriate that understanding to deliver new drugs.

> Madagascar's unique rose periwinkle plant was used to develop two anticancer drugs, vincristine and vinblastine, which together have generated more than $100 million in sales for a global pharmaceutical company. Madagascar, however, got no financial return from these discoveries.
> —*World Development Report 1998–1999,* p. 35

Very little has been done in terms of the broader understanding of ecological processes and local ecological histories and using that knowledge effectively in developing good ecological management.

> While economic indicators such as investment, production and trade are consistently positive, the key environmental indicators are increasingly negative. Forests are shrinking, water tables are falling, soils are eroding, wetlands are disappearing,

fisheries are collapsing, rangelands are deteriorating, rivers are running dry, temperatures are rising, coral reefs are dying, and plant and animal species are disappearing. The global economy as now structured cannot continue to expand much longer if the ecosystem on which it depends continues to deteriorate at the current rate.

—Lester R. Brown, *State of the World 1998*, p. 5

Devolution of Power and Ecological Management

In India there has been slow but steady progress in empowering local villages or tribal hamlet communities to make decisions. I am very excited about that. In many parts of India I have worked with local and tribal village councils. I have talked with them about their perceptions and how they see local natural resources being managed and how they wish to deal with the power that they have under the newer legislation in many parts of India.

I have formulated an inverse law that the concern of the politician with environmental issues is inversely proportionate to the size of his or her constituency. Members of Parliament are very little concerned with what happens at the local village levels, but the people elected to the local village councils are very much concerned. As the power has slowly devolved to the lower levels, we see all over India the beginning of a shift. Among the best experiments have been the village forest councils. Where they are effective the devolution of power to the local level has maximally progressed. They are the best examples of good management and often a good rehabilitation of the local forest ecosystems.

> Little of the economic benefits from forest exploitation return to the communities who lost access to forest resources. In fact, their standard of living has declined. Most of the profits benefit a few powerful industries or families. The liquidation of 90 percent of the Philippines' primary forest during the Marcos regime, for instance, made a few hundred families $42 billion richer, but impoverished 18 million forest dwellers.
>
> —Janet N. Abramovitz, *State of the World 1998*, p. 29

Local Wisdom

I believe that local communities are still highly motivated to maintain their local environment in good health. They are dependent on a very wide diversity of biological resources to fulfill many of their requirements, including earning money by the sale of medicinal plants.

> One of the best-known contracts (providing compensation for the exploitation of indigenous knowledge and bio-resources) is that negotiated between Mercks

& Company and INBio, Costa Rica's nonprofit national biodiversity institute. Merck provided $1.1 million initially, plus a commitment to share royalties on any commercial products developed, in exchange for 20,000 to 10,000 extracts from plants, insects, and microorganisms in Costa Rica. INBio has now entered into nine research agreements giving companies limited access to biological resources in return for financial compensation and technology transfer.

—*World Development Report 1998–1999*, p. 35

I think that if we allow local people to make decisions, then they would make decisions that are largely environment-friendly. Certainly they may need help in arriving at and implementing those decisions; nevertheless the local people must have the most important say. The decisions have to be flexible, specifically geared to local conditions and local ecological wisdom. Certainly we need to take into account traditions such as the protection of nature by Bishnois and the many other sacred forests, sacred ponds, or sacred trees like the Ficus trees.

Local Democracy

The difficulty is that when local people take over their own destinies, then the biosphere people, who today hold the power from cities and the capitol, will certainly have to sacrifice something. An example would be that they will have to pay real prices for natural resources, because we get a large number of resources at subsidized rates. For instance, the paper industry gets bamboo from state controlled forests at rates that are often substantially lower than rates the local basket weavers pay.

There is a reluctance to share power so also there is a reluctance to share knowledge. There hasn't been enough investment in education, for instance. But all this is changing because democracy has deep roots in Indian society and people are demanding that they are better educated. They want a greater share in decision making. They want control over their own resources and there is a slow and steady shift in that direction. I believe this would, in maybe another ten, twenty years, lead to a situation where there will be a much greater genuine democracy and a much greater local involvement in managing the environment. I am quite sure that there will then be a situation far more favorable to good management of the country's resources.

Consumption of paper (including newspaper and paperboard) is increasing faster than any other forest product. The world uses more than five times as much paper today as it did in 1950, and consumption is expected to double again by 2010. About two thirds of the paper produced worldwide is made form virgin logs; only 4 percent is made from non-wood source such as cotton or rice straw.

The rest comes from wastepaper. Soon paper production is expected to account for more than half of the global industrial wood harvest.
—Janet N. Abramovitz, *State of the World 1998,* p. 29

Ecology Is Different from the Other Sciences

Ecology is different because ecology is dealing with historical complex systems that cannot be manipulated at will at all. Scientists are beginning to realize that when we are dealing with complex systems such as ecological systems, that science remains at a stage that is not very much advanced over folk science. Therefore we must take people seriously and listen to them. I think this is a very healthy sign for science to shed its arrogance, to accept a certain modesty, and to really bring in a genuine democratic culture, a much more participatory effort.

When you come to ecological issues the technical people must work hand in hand with people who have deep understanding based on their own experiences of these complex systems. Perhaps we will develop a notion of a new culture of science when dealing with our complex issues of environment.

In that sense, forest-produce gatherers and basket weavers are all practical ecologists and unless we work hand in hand with them, we are missing an enormous store of information that could be brought to understanding, especially of local ecologies and prudent management of local ecological resources.

Science and Traditional Wisdom

The principle of evolution through natural selection is an enormous advancement in our understanding of the natural world, the way it is shaped and functions. That is obviously not part of folk understanding. Nevertheless, this understanding that gives us such deep and beautiful insights into the natural world has a very little role to play in terms of actually managing this natural world because we are managing it on scales that are very different from evolutionary timescales. Frankly, I do not see science having advanced in terms of our understanding of managing it on these short time scales.

In the Indian epic the *Mahabharata,* which is perhaps three thousand years old, there is one particular episode, where the heroes, the Pandavas were hunting in a forest. The local animals appeared into the dream of the eldest of the Pandavas and told him, you are over harvesting our population. The next morning they decided to go to another part of the forest.

In the Andaman Islands they have a tradition that once a pig has been hunted in one patch of the forest, they mark that patch and they will move to another patch and come back at least three weeks or more later. That is the notion that you must rotate your harvesting pressure, which is also part of modern ecology. Though I try very hard as a professional ecologist, I cannot come up with something ecologists can tell that is entirely new. I think it is a very healthy sign for science to shed its arrogance. We need to accept a certain modesty and to really bring in a genuine democratic culture, a much more participatory effort in certain kinds of scientific endeavors and certainly when we are dealing with ecology.

Love of Life

I think Indian culture has a tremendous expression of the love of life. There is a whole range of organisms whom we venerate and protect even today. There is a pond dedicated to one of the Muslim saints that harbors the only surviving population lives of the turtle species Trionyx nigricans, and all over India roam macaques. We enjoy watching them because we see them as our kin. We have a feeling of being part of a community of beings.

Most Indians believe quite seriously in rebirth. You could say that how many births you go through is an estimate of how many kinds of different organisms there are. I am very intrigued by the fact that according to Indian cosmology there are 8.4 million different kinds of creatures in the entire chain a human being passes through in his or her reincarnations.

Now the modern estimate of the total diversity of species has fluctuated. When I was a graduate student it was three million. Then it shot up to 30, 50 million, but now people are talking about 8 to 12 million. Our 8.4 million is exactly in that bracket. Maybe we have a genuine feeling for being part of a chain of beings of this great level of diversity.

My Hope for the Future

As far as we know, humans are the only prudent species of animal. There is no other animal that thinks of the consequences of their actions in the long term and that takes appropriate measures. Humans have shown that they can and do take measures to conserve nature, throughout the historical period. This has been patchy and perhaps we have destroyed nature over human history much more often than we have taken positive measures, but again and again, humans have taken very positive measures.

In the great famines of the seventeenth century, the Rajasthani vegetation had almost been destroyed and the Bishnois came up as a new religious sect that brought back the greenery and the animal populations. In Europe we know there had been very extensive deforestation but countries like Switzerland now have a very healthy forest cover. The forest protection committees of West Bengal have brought back the natural forest and their animal populations.

All over the world there are examples and there are continuing examples even in India today. I think there are very definite signs that eventually we will roll back the tide and we will move towards a greener and environmentally sounder future.

> A disproportionate share of the world's industrial roundwood is consumed in industrial nations. In fact, 77 percent of the world's timber harvested for industrial purposes is used by the 22 percent of the world's population who live in these nations. Although developing nations have been increasing their share of consumption in recent decades, they are still well below the levels found elsewhere. Indeed, consumption per person in industrial nations is 12 times higher than in developing ones. Fuelwood is the only wood product that developing nations use more of, and even then their consumption per person is less than twice that in industrial nations despite the fact that it is the dominant industrial and household energy source in some developing nations.
>
> —Janet N. Abramovitz, *Vital Signs 1999*, p. 76

References

Gadgil, Madhav and Ramachandra Guha. 1992. *This Fissured Land: An Ecological History of India*. Oxford: Oxford University Press.

————. 1995. *Ecology and Equity: The Use and Abuse of Nature in Contemporary India*. London: Routledge.

II

Science, Policy, and the Public

FEDERICO MAYOR

The Eyes of the Universe

Federico Mayor

Born in Spain, trained in Pharmacy, Dr. Federico Mayor has been Professor of Biochemistry at the University of Granada and at the Autonomous University of Madrid. He was Director of the Department of Biochemistry in Granada, and of the Molecular Biology Centre in Madrid. In the public arena, Dr. Mayor has held posts as Minister of Education and Science in Spain (1981–1982), Member of the European Parliament (1987), and Deputy Director-General of UNESCO (1978–1981). Recently he was Director-General of UNESCO from 1987 to 1999. Under his Directorate, the *World Conference on Science—Science for the Twenty-first Century* was held in Budapest in June/July 1999. He is a member of numerous national science academies and has received several honorary doctorates from various universities.

Dr. Mayor, you play a major role in this UNESCO/ICSU gathering in Budapest. What are your many hopes this week—not only in terms of the level and scope of scientific dialogue, but in terms of how the world's citizens, and the world itself, may benefit?

What I expect is a public awareness that the present trends must be redressed. The world is not going well. Economically the gap is widening instead of narrowing. In the environmental aspects things are not going well. Also in the social aspects, particularly in the last few years there has been a failure of the mechanisms for development. We thought that we would give loans and that these loans will support the endogenous development of the countries so they themselves can exploit their natural resources. This has not been the case. They are indebted and there has been external assistance in such a way that again the countries that have the technology and that have the financial resources are going up and the others are going down.

The same in my view holds for the cultural aspects. We have at this moment a universal utilization of the media that is very favorable, which we can utilize for wider education, but at the same time they can produce a progressive uniformization of cultural diversity.

There is an immense asymmetry in the gender issue. Today we live in an androcentric society. Ninety-six percent of decisions at the worldwide level are taken by men. In parliaments for example, men represent 90 percent of the voice of the earth.

The Role of Scientists

My view is that the voice of the scientific community must be louder and louder. We must say, enough is enough. During this century we have made the most fantastic discoveries. Antibiotics, sophisticated surgical procedures; we can deal with diseases like never before. We are able to communicate with each other using portables; we can go online, whatever you like. All this is to the benefit of humanity. At the same time this has produced this immense asymmetry between the haves and the have-nots. An immense asymmetry also in the sharing of knowledge.

We have been in our ivory towers. Now we must say that we are going to raise our voice, not be silent anymore and say things like they are. We must provide the decisionmakers with good elements for decisions to which they must listen. If they don't listen we will raise our voice. This is democracy, the voice of people. I think that in a non-violent way we will be able to change the present trends.

> Science bears a general responsibility for the well-being of humanity. Every scientist must have a constant awareness of the possible consequences of her/his, research. A full and open dialogue involving various sectors of society is necessary to consider the consequences of experiments like human cloning and genetic engineering before their initiation. More forceful legal and moral safeguards should be worked out to prevent unethical practice and misuse of science for the development of mass destruction weapons, and for experiments which disregard the dignity of human persons or animals. A special education on ethics should be included in all education curricula. We expect and encourage the global scientific community to try to find a consensus about the self-regulation of science. We strongly support the establishment of a scientific Hippocratic oath.
>
> Statement, International Forum of Young Scientists, Budapest, 23–24 June 1999

The Power of Non-Violence and Democracy

I am aware of the immense force of people. How can you really counteract power, force, and imposition? Look at the historical examples, Martin Luther King, the Mahatma Gandhi; they were non-violent. They were able to persuade millions of people. Very recently UNESCO has published *The Power of Non-violent Action* that I gave to Corretta King in Atlanta.

The media are crucial. Without the media we cannot achieve this mobilization of will. The last century, in which we have done so many wonderful things, has been a century of war and violence, a century of the turmoil of humiliations and vengeance. We must learn to live together peacefully and we must learn to be persevering in our disagreement without utilizing force and violence.

National and International Democracy

Take, for example, what has happened in Kosovo. In my view we must learn that there are democracies at the national level and there is democracy at the international level. This democracy has a name. It is the UN. If some countries consider that the United Nations must be reinforced, if we are too weak, please give us moral support. If we are too slow, please accelerate our procedures. But if you act outside the UN system then the precedent is terrible, because other countries can do the same. If you break completely out of the international legal frame at this moment the consequences can be very important. Democracy is the solution for all, even for poverty. We want to have the voice of the people and democracy at the national level. But we cannot accept—because it is incoherent—an oligarchy at the international level! That is unacceptable.

How Can Democracy and Poverty Be Reconciled

Democratic India is a miracle, because don't forget that there are nearly one billion people. In general there is no violence and they are within a democratic frame. China is very closed; they count one billion three hundred million, but they are not in a democracy. When I wake up in the morning I say: "How are things going in India? If they are going well, then there is still hope."

Imagine that in India there would be this Mahatma Gandhi Way of Action, without violence, which is so powerful. They were a country under an immense need for nutrition until recently. Today they are exporting grain. M. S. Swaminathan is one of the key persons, one of the three or four of the Green Revolution. They have demonstrated that you are able to live frugally, sometimes in poverty, but not in misery. This is an important distinction. They are clean; you can come and meet them in their houses. They can be illiterate, but they are very wise.

In the arrogant Western European countries we sometimes say that they are illiterate, and we think immediately that they are ignorant. I have found a letter addressed by Mahatma Gandhi to Julian Huxley, who was UNESCO's first Director-General, and who was very clever, putting things on the tracks. This was in February 1947. Mahatma Gandhi said: "I have consulted the draft document that you have sent to me (on the Declaration on Human Rights, by the way) with the most intelligent person that I have ever met. A woman, illiterate—my mother." This is to say, in Africa, for example, I have met so many people who are very poor, but they are so wise. It is for this that I say, well, we have the know-how, but they have the wisdom. This quality of human beings is the first article of the Declaration of Human Rights. Sometimes we forget that we are all equal in dignity. If we are all equal in dignity, then all problems are solved. Including all the problems of a country like India.

Peace, Democracy, Development—And Justice

To have sustainable peace we must have sustainable endogenous development. Development means sharing, justice, and freedom. I am so convinced that only if there is freedom of expression, we will have the only tool to counteract violence. Including the violence from the state, the imposition from the state, or from some states. Only through freedom of expression can we change legality into justice.

What we need is not "the rule of law." In Hitler's Germany there was the rule of law, and also with Stalin. We need the rule of justice. For the rule of justice we need newspapers, television, and the radio. We need the people passing their message, including the young people. The media should now be more available for the young people to express themselves freely. Peace is a precondition for development, a precondition for justice and for freedom, for democracy.

This triangle of peace, democracy, and development is so important. How can a young man or girl contribute to this movement? I think that is very important because only through the media we can imagine that they can express themselves. Or through associations, as I mentioned, the UNESCO Clubs and the Association of Young Scientists.

> It is plain to see that no nation is entirely exempt from a crisis of values, or perhaps one should say a moral crisis. And yet . . . Humanity possesses the knowledge and resources with which it could put an end to most of these crisis, to eliminate, reduce or at least considerably dilute the causes of these present instances of appalling inequality, injustice, discrimination, exclusion, frustration and humiliation. On emerging from "the great and terrible war" it provided itself with a set of mechanisms for concerted action such as would enable the nations of the world and men and women of goodwill to work together for world peace and the common welfare and well-being of all peoples. Not the least among those mechanisms is UNESCO itself.
>
> —Federico Mayor, *UNESCO, An Ideal in Action,* p. 10

If the size of the public and particularly the private supranational conglomerates is increasing very fast (mega-fusion processes), the NGOs, and the academic and scientific institutions should associate in a worldwide network, in order to be able, all different but all united to this end, to have the necessary stature to be listened to and to be taken into account.

We Are All Citizens of the World

The first victim of war is truth. All sides put things in favor of their respective actions. War is perverse. I am very afraid that whenever we speak through the news in favor of one culture or one people we are speaking against the other. Children, women, and ordinary people have no nationality. For me all are citizens of the world. Now it appears that for many people the Serbs are not good.

And the others are the good people. That is wrong. The Iraqi children are wonderful. The Iraqi women are excellent. The ordinary people of Iraq are wonderful too. The same is true for the Serbs and the Kosovars, for everybody. People are basically good.

Yesterday they came to me, saying, how can we accept that this rapporteur is from this country? I said scientists have no nationality. It is unacceptable that we have these kind of clichés. This is promoted by the lack of good independent information.

What Is UNESCO?

UNESCO is the Organization of the United Nations System whose goal is to promote peace and to build peace through Education, Science, and Culture. During the Second European Great War, there was a minister of the United Kingdom, Butler. He said already in 1942, that all these terrible things happen at this moment because we have not been well educated. There was a poem of Archibald McLeish, a poet from the United States, that went, "Since wars begin in the minds of men, it's in the minds of men the defences of peace must be built." Butler said: "After this war is finished, we must educate the people and build peace through education."

> What is lacking is the determination to overcome national self-interest and short-term concerns and to set out upon the road of solidarity, the courage to proclaim and to convince others that, in an increasingly interdependent world, peace and prosperity are unsure and fragile unless shared by all. This determination is something that has to be created. the United Nations system bears a great responsibility in that respect . . . By virtue of its acknowledged ethical mission, UNESCO has a pioneering role to play in this undertaking.
> —Federico Mayor, *UNESCO, An Ideal in Action*, p. 11

He had a meeting of the shadow Ministers of Education of Europe. At this meeting they already thought that culture should be added. The UN did not yet exist. Then, at the end of the war, there was the meeting in Bretton Woods about the economic and financial aspects (the World Bank and the International Monetary Fund). It was at the meeting in San Francisco when the UN was created. And the UN Charter says: "We the people, we have decided to save the succeeding generations from the horrors of war." Then they realized the need of an intellectual arm, because the UN is a political space and now we need the intellectual institution, what Jawaharlal Nehru called "the conscience of the system."

Archibald McLeish was very active. In fact he was one of the writers of the wonderful text that is UNESCO's Constitution. In London Winston Churchill and Richard A. Butler were no longer in the government, but Clement Atlee and Ellie Wilkinson, the Minister of Education, were in the government. Atlee was also one of the drafters and enthusiastically he took this phrase of Archibald McLeish. Through education and culture we must build peace in the minds of men! Professor Joseph Needham, a Sinologist and Mathematician from the United Kingdom clained: "How can we achieve this without scientific knowledge?" That was the moment when the organization was created to build peace in the minds of men and women through education, science, and culture.

UNESCO's constitution was adopted by the London Conference in November 1945, and entered into effect on the 4th of November 1946 when 20 states had deposited instruments of acceptance. It has 188 Member States as of 19 October 1999. The main objective of UNESCO is to contribute to peace and security in the world by promoting collaboration among nations through education, science, culture and communication in order to further universal respect for justice, for the rule of law and for the human rights and fundamental freedoms which are affirmed for the peoples of the world, without distinction of race, sex, language or religion, by the Charter of the United Nations.

—UNESCO web site, 2000

Freedom of Expression Is Crucial

According to its mission, the drafters of the Constitution considered the freedom of expression essential. Therefore, Article 1 is as follows: "UNESCO must guarantee the free flow of ideas through word and image." At the beginning of the eighties, it was suggested to add, after the McBride Commission: "free and balanced." This balance could be terrible! At that time I was Minister of Education and Science in Spain, and I rushed to Paris to urge, that no change should be introduced and unrestricted freedom preserved.

The United States was already under an administration that was against the UN, not against UNESCO. It was even said: we would very much like to waive the UN out of Manhattan. They probably thought that the most vulnerable organization of the UN System was UNESCO, because we are not a Fund and not a program with a lot of money. Perhaps they thought that intellectuals are more dangerous, and they withdrew from UNESCO.

They were wrong, because you can kill an institution, but you cannot kill an idea. You can kill a messenger, but you cannot kill the idea that he brings. At that time the United States paid 25 % of the budget and the United Kingdom 6 % of the budget, so together that was 31 %. Despite this reduction, nothing has happened. Today UNESCO is stronger than in the past. We have international support through extrabudgetary funds that are more than four times the contribution of the United States.

As of 31 December 1998, UN member states owed a total of $1.6 billion for peace-keeping operations, or roughly two years' worth of operations. The United States remained by far the largest debtor, with $976 million in unpaid dues—61 percent of the total.

—Michael Renner, *Vital Signs 1999*, p. 114

The United Nations, UNESCO, and the United States

What does this mean? It is a good lesson. If you consider that the United Nations is not doing well, then you must try to do better, and you must give the resources, the staff, and the possibilities to do better. In recent years, for example, the United States has not signed the Law of the Sea, the Convention on the Children, or the Anti-Person Mines Protocol. You realize that in fact this is not a position against UNESCO but a kind of lack of confidence in the UN.

I would very much like to persuade the United States, that the UN is the only framework that we have today at the international level for legal and ethical guidelines. Of course there must be a change in the composition, in the functions, but the United States cannot be absent. How can we imagine that an international convention, for example, the Kyoto Agreement on CO_2, can be functional if the United States has not signed?

In relation with UNESCO, the United States signed up to Man and the Biosphere, to the Convention on World Heritage, the Copyright Convention, and they participate now in other activities as well. For me it is as if the people of the United States are a member-state. We cannot live without the people of the United States, their scientists and educators. I am so much indebted to the U.S. National Academy and to the teacher's trade union, for example. Today there is only one, but at that time there were two in the world, the Soviet and the non-Soviet one. Today there is only one teachers' trade union, led by the wonderful American woman Mary Futrell. With the American Association for the Advancement of Science we have had excellent relations as well.

One thing is very important: the appointment of staff. On a particular issue I was told that an American was the best in the world. I appointed the American. I have done this because I consider relations at the state level one thing, but relations at the people's level is another. We have said in the charter: "We the peoples," not "We the governments" or "We the states."

In November 1995, on the occasion of the fiftieth anniversary of UNESCO, President Clinton addressed a letter to our General Conference. He said that his administration had decided to rejoin UNESCO. However, the United States cannot act in this way right now because of the problem in Congress with this difficult cohabitation of the Democrats and the Republicans. It is the willingness of President Clinton and his administration to rejoin, but there are the problems with the budget.

UNESCO is an organization of intellectuals, scientists, artists, writers, and journalists. We can live very well with the support of other countries and with the widespread distribution of our ideals. One idea is more important than many dollars. I think we have in UNESCO some ideas that can mobilize the world better than many millions of dollars.

UNESCO and Environmental Issues

Many years ago the biologist Julian Huxley, UNESCO's first Director-General, created the International Union for the Conservation of Nature. This was in 1947, 1948, and it can be said it was a "daughter" of UNESCO. Huxley was a pioneer of the preoccupation with environmental issues. Today we have more than five hundred areas in the world that are already under the "Man and the Biosphere" project, which means that we can guarantee the good quality of environment in those areas.

How Can Young People Be Involved with UNESCO?

There has been a Meeting of Young Scientists just before the World Conference in Budapest. Very important personalities were there such as M. S. Swaminathan, Leon M. Lederman, and Julia Marton-Lefèvre. It has been a significant consultation, also at the regional level. I think in different countries we have made good encounters with the young people through the UNESCO Clubs, of which we have more than five thousand. For teenagers we have the Associated Schools.

The year 2000 was the International Year for the Culture of Peace. For me that is the most important prerequisite for development and democracy. I have been in Bosnia-Herzegovina. I have been in countries where three years ago there was the turmoil of ethnic clashes. What is the message that young people give us? They say: when there is violence, there is no education, no justice, no development. We must be very attentive to their unanimous voice: we do not want to pay the price of war anymore.

> The year 1998 marked the 50th year of UN peacekeeping. Only 13 operations were established during the first 40 years compared with 36 in the past decade alone. More than 750.000 persons from 110 countries have served in all UN peacekeeping operations. Of the 49 missions to date, 17 were dispatched to countries in Africa, 9 each to Europe and the Middle East, 8 to Central America and the Caribbean, and 7 to Asia.
>
> —Michael Renner, *Vital Signs 1999*, p. 114

Federico Mayor

My Hope for the Twenty-first Century: The Eyes of the Universe

My hope is in education, education, education. Education means much more than information or instruction. Education means to give to each human being, women and men, personal sovereignty. You decide because you reflect and you say yes or no. Not because some sect or association tells you what to do. You decide *by yourself.* You do not decide because you are hungry or because perhaps you will not survive otherwise. Access to education for all human beings will give the same dignity to all of them. That is article number one of the Declaration of Human Rights. If all human beings can decide by themselves, then the world will be the world that we wish. That is what we expect and hope for our children. Otherwise what happens is that a very large number of people will not have not the opportunity of becoming masters of themselves.

I tell you this as a biochemist. Today we can forecast the behaviors of insects and animals; we know the hormones that activate their receptors. It is all measurable. There is only one living being that is not measurable. The human being. That is because we are able to create. Each human being is unique, and is a wonder. Able to create and to go beyond his or her limited biological structures. We can produce wonderful music and ideas. . . . All human beings are miracles. At least, they are a mystery.

Sometimes it is said that UNESCO is excellent because we have safeguarded sacred sites such as the Pyramids, Machu Picchu, and so on. All these are stone monuments; they are not vulnerable. They cannot create. Human beings are the eyes of the universe, and these eyes that know what is happening and what they expect are our hope.

Every day we must sow the seeds of a better future. I admit that many of the seeds we plant will not bear fruit. But it's only one fruit we will never collect: the fruit of the seeds we didn't have the courage to plant.

References

Mayor, Federico, in collaboration with Sema Tanguiane. 1997. *UNESCO—An Ideal in Action. The Continuing Relevance of a Visionary Text.* Paris: UNESCO.

JOHN DURANT

Hope, Anxiety, and Doubt:
The Public View of Science

John Durant

John Durant is Professor of Public Understanding of Science at Imperial College in London. He was Assistant Director and Head of Science Communication at the Science Museum in London, where he was also responsible for the construction of a new wing dedicated to contemporary science and technology. Since 2000 he has been Chief Executive of the science center *at-Bristol*. Widely published and a regular contributor to radio and television programs, Professor Durant is also Founding Editor of the quarterly international journal, "Public Understanding of Science." From 1999 to 2001 he coordinated a major European Commission-funded international research project on "European Debates on Biotechnology: Dimensions of Public Concern."

Professor Durant, you have devoted yourself to education in the sciences. Tell us what it is that motivates you, and clarify some of the burning issues that seem to be of greatest concern to the public.

People are probably more interested now in science and technology than ever before. At the moment in many parts of the world, there is quite a boom in public interest in science. There's a fascination with it in the form of popular books written by leading scientists. Most of that literature is in areas like cosmology, evolutionary theory, and some of the remoter fringes of brain science. These are not being sold to people because they're going to change their daily lives. What you're going to get is a greater insight into the universe you live in, and your place in it; a kind of age-old philosophical appeal.

> Why do Americans, and to a lesser extent Canadians, support science so strongly? Paradoxically, their much maligned education systems may be part of the answer. Americans attend college in greater proportions than young people in most other countries. In college, the core curriculum includes science courses. This exposure may be minimal but it exceeds that of the more specialized education systems of other countries.
> —*UNESCO World Science Report 1998*, p. 47

There is however a mixture of attitudes. On the one hand people are assuming that the world is being somehow made potentially a better place through science. Alongside that sit a whole series of anxieties, worries, and doubts about problems that increasingly seem to have a scientific twist to them.

> The promise of science is twofold: deeper understanding and material benefit. The world at large reflects the impatience of the research community. When there is a promise that a certain kind of cancer will be cured, it is natural that people suffering from that condition should be angered by delay. Or that people in developing countries should be impatient that the prosperity they see elsewhere, much of it the technology born of scientific discovery, is not more quickly transferred to them.
> —John Maddox, *UNESCO World Science Report 1998*, p. 17

For example, in the area of genetics there is an endless series of difficulties. Practical ones about whether we want to eat genetically modified foods, moral problems about how we use genetics in health care, cloning, those kinds of things. People are fascinated and they have high expectations and some anxieties.

The Debate about Biotechnology

The debate about genetically modified plants for food has been brewing in Europe in general for years. In the United Kingdom at least since 1996 there has been a growing sense of discontent especially about food. We've had ten or more years of worries about BSE, the so-called "mad cow disease," where the best scientific advice way back has proved actually not to have been right.

The problem was with a new kind of disease, a so-called encephalopathy that affected cattle. Nobody was sure where it had come from and whether it could somehow pass on to people. Then a new form of encephalopathy in people was diagnosed and a small number of deaths have taken place. The best guess is that they are probably due to this, but it's a very hard disease to research, and it sensitizes people. They wonder, "What are they doing to our food now?"

When the first genetically modified foods came through they were specifically labeled, tomato paste, for example. From 1996 on we had soya and maize coming onto the market but these were unsegregated so you couldn't distinguish the genetically modified and the unmodified. GM (genetically modified) soya mostly benefits farmers in growing them, not the consumers who eat them, as far as I can tell. Some estimates say that up to 60 percent of all the processed foods in a supermarket might combine some of these genetically modified commodity crops. Now people began to say, "Hang on a minute; I don't get a choice here? Can I not choose whether to buy this or not? What good is it for me?" A whole series of questions began to be raised in the public domain.

> Genetic engineering is often justified as a humane technology, one that feeds more people with better food. Nothing could be further from the truth. With very few exceptions, the whole point of genetic engineering is to increase the sales of chemicals and bio-engineered products to dependent farmers.
> —David Ehrenfeld, "A Cruel Agriculture," in *Resurgence*, March/April 1998

The Pusztai Case

The lid blew off when a Hungarian scientist called Arpad Pusztai who happened to be working in Scotland broke a story of some unpublished research that he was doing. This seemed to show, I have to stress it *seemed* to show to him, that genetically modified potatoes fed to rats might cause adverse health effects in those rats. That work has not been substantiated. This is an unfortunate case of somebody going public with science before it really had been authenticated. It blew the lid right off the GM food debate and this led to the most extraordinarily intense debate.

I don't know of any authentication of his findings and the best available reports I've seen have suggested that there's nothing in it. I don't think that there's any substantiated health risk to eating GM foods but there are a whole series of other questions being raised having to do with consumer choice and environmental effects.

An Uncontrolled Experiment with Nature

Perhaps the most difficult question for the GM industry is, "What about cross-pollination from GM crops in the field to non-GM ones?" We've had a boom in interest in organic food recently in some parts of Europe, including the United Kingdom. The organic farmers are saying they don't want to grow GM and what's more they want to be sure that their crops are not being contaminated. These are very complicated questions but they're typical in a way of some of the difficulties that science and technology can get into these days. It's not just a story of unalloyed progress and everything being bright and beautiful.

> The United States has dominated the explosive growth in genetically modified crops, with 74 percent of global acreage. Argentina and Canada trail a distant second and third. Only in these three nations do transgenics constitute a substantial share of any crop harvest. In 1998, more than a third of the American soybean crop was transgenic, as was nearly a quarter of the corn and 20 percent of the cotton. Roughly 55 percent of the Argentine soybean was transgenic, as well as 45 percent of the Canadian crop of rapeseed, which is crushed to produce canola oil.
> —Brian Halweil, *Vital Signs 1999*, p. 122

The whole of modern agriculture worldwide is a completely uncontrolled experiment with nature. The irony here is enormous. The whole of modern agriculture is based upon selective breeding of plants and animals, which has been going on for hundreds of years, and in recent decades intensively. Nobody has done any checks whatever, as far as I'm aware, at any stage in that process to find out what effects on the environment these new varieties have. It is peculiar that we're taking one particular new way of modifying plants and animals and asking it to go through a bunch of hoops that all the other methods have never had to go through. I'm not saying that it is wrong, but we need to get this in context.

The Disappearance of Nature

You look around any part of Western Europe today and you will not find nature. There is no countryside or land that is unaffected in its ecology by human cultivation or human management. Every single square yard of the United Kingdom is intensively managed land, even on mountaintops. The question then is, are we being realistic? We have had enormous effects on our ecology already. GM at the moment is a very tiny additional step.

What's interesting is that a much bigger debate is being opened up in Europe about what we want to do with respect to our relationship to the countryside, with respect to our relationship to wider ecosystems. Do we really want to just change them purely to produce as much food as we can as cheaply as we can? Do we want to preserve a certain amount of biodiversity in our countryside; do we want our farmers to be responsible for looking after nature for us, as well as looking after food supply? That's a debate we haven't had, certainly in my country, since the Second World War when there was a food shortage and the message to farmers was, "You grow as much as you can, as fast as you can; we won't ask any questions." Now we're asking questions.

How to Preserve Semi-wild Land

We have to remember that the word biodiversity is only some fifteen or eighteen years old. The focus for biodiversity was, quite rightly, on world biodiversity and on the parts of the world where you have enormous biodiversity. Many of them were threatened habitats—parts of the tropics, the rain forests, the seas, and so on. In recent years the focus has come, in a sense, back home for some of us in the Western industrialized world at least. What do we really think about the biodiversity on our doorsteps? Not just the biodiversity in those exotic places you see in travel movies. That is where some of the GM debate is having its effect.

The Public Attitude towards Biotechnology

For some years I've been researching the way in which European public attitudes to biotechnology have been moving and we have compared those with similar studies in Canada and the United States. Interestingly you see a very large range of views about different biotechnologies. There is no single view among the public about what we're doing with gene technology these days. People tend to view plant agricultural technologies quite differently from say, what's going on in industry with the creation of new enzymes using genetically modified micro-organisms. They view that quite differently from the way that gene technology is being used in medicine, to produce new drugs, new vaccines or genetic tests.

> The US and Canadian publics' assessment of the benefits and risks of research show great confidence in science and technology. Across many countries surveyed regarding both positive and negative attitudes to science and technology, the U.S. public is by far the most positive. The Canadian public comes second, people in Japan are the most pessimistic and those in Europe fall in between.
> —*UNESCO World Science Report 1998*, p. 47

Interestingly, in recent years the views in Europe and the United States on the nonagricultural biotechnologies have become very similar. Europeans are even more in favor of new gene technologies in some areas of medicine than U.S. citizens. It's when you get to food and agricultural issues that the Europeans are much more wary.

Some of us are wondering when the debate is going to break in the United States. I've been in some U.S. supermarkets and there seems to be quite a market for so-called natural, healthy and organic food. That is part of the impetus that lies behind the opposition to genetically modified foods in Europe, but it doesn't seem to have done that in the States. Maybe this is a situation where the United States is running a little behind Europe for once.

The European Debate

I've already mentioned some of these anxieties about new food technologies that are born of the BSE crisis. That is very hard to underestimate. Billions of euros have been lost by the beef livestock industry in Europe. It has been a complete economic and public relations disaster, and it could still be a human health disaster. This has been going on for more than ten years. There's been no BSE reported to date in the United States which signals an important difference between Europe and the United States.

I also think there are agricultural differences. There is no wilderness in Western Europe. In the United States, very fortunately and wisely perhaps, people set aside huge amounts of land in the early days of the American Union that was not to be used for anything except wilderness. There are very strict regulations about what people can do in those regions. I sometimes wonder whether the environmental concerns of many U.S. citizens don't tend to focus on these wildernesses and their preservation, and perhaps not so much on what's going on in the great wheat fields of Iowa, for example. In Europe we don't have that luxury. We have to get our conservation and our natural history if you like out of our agricultural land because that's all we've got left, by and large.

Animal Rights

There are other issues. It's interesting how many of them at the moment seem to be in the life science area. Animal rights certainly are a long-standing concern. Again it's international. In the United Kingdom, probably in the last ten, twenty years, animal rights have been the single most polarized issue.

I live in a town in Oxfordshire that happens to have just outside it the only cat farm in the country that is legally breeding cats for use in medical research. The farm has become the object of sustained high-level public protest. You'll have thousands of people congregating on the town to march and picket outside this farm. Helicopters overhead, cameras on high podia, hundreds of police out in uniform. And this has been going on now for years.

According to public opinion polls most people actually support the careful use of animals for good reason in medical research. But there is a minority who passionately opposes it. I suspect that if you took a poll among 15-, 16-year-old teenagers in Britain you might get a majority who were against animals in research. I think we have to realize that there are very difficult moral issues here with respect to what we're prepared to do to other animals for the sake of our own health and well-being.

Major Concerns about Cloning

What interests me about cloning is that this is one of those rather unusual examples, I'm thinking particularly of Dolly the sheep, where you get virtually a worldwide, instantaneous concern. Within thirty-six hours of the press conference announcing that Dolly the sheep existed, you had governments on several continents issuing statements, including the United States and half the nations of Western Europe.

This was an extraordinary level of international concern about the fact that actually we contrived in a slightly unusual way to produce two sheep that were genetically identical. You could say that this is ironic because most people regard sheep as identical anyway; it's very hard to tell two sheep apart.

Again you can see cloning as a carrier for a whole series of other anxieties that have to do with what we're doing to our understanding of life and to our ability to influence life and death with science. People talk about xeroxing life. This is a very uncomfortable idea that we could produce identical copies of our chosen animal, even though we've been doing it with plants for years. The thought that we could do that with people is, of course, what really triggered the big public debate.

Human cloning has become a subject of intense public interest and public concern and most of the evidence we have suggests that most people do not like that idea. In the United Kingdom the government has actually decided to pause, even over the use of cloning for purely research or therapeutic purposes, even though scientists were saying you could get new treatments out of this.

Science and Industry

The economic reality is changing; far more science is funded by industry and for industry today than, say, one hundred years ago and we need to understand that. A far larger proportion of scientists are working for people who have commercial interests and, again, I think all the evidence is that the public are canny about this.

The first question that people ask about scientists these days is, "Who do they work for?" If the answer is, "They work for a university where they're doing publicly funded research," then that will be one way of interpreting what they say next. If the answer is, "They are working for a multinational corporation" or even for the government there's another. You don't take people just as scientists; you take them as people situated in a particular institutional setting and you ask the question: "What's the spin?" Science has got to get used to that attitude to itself.

Use and Abuse of Science

One of the turning points for science and for its sense of its own responsibilities in the whole of human history was, of course, the chain of events at the end of the Second World War when the atom bomb was created. For many physicists that was a life defining moment. Some physicists left the subject and were never to go back. Some became ethically and politically committed from then on to work in favor of science for peace. Joseph Rotblat, who won the Nobel Prize for Peace is one of those physical scientists whose life was changed by seeing his specialty become involved in creating the world's most fearsome weapon of mass destruction. Ever since he's been campaigning for peace and for "science for peace" and for higher ethical standards in science. I don't know anybody actually in science, who doesn't admire him and his stance.

When you look at the way in which science works in the world, one of the difficulties is that actually its effects are so varied and unpredictable. It's very hard to be sure that any particular piece of science doesn't have the potential for misuse. There's a very poignant story from a little later in our century. An American plant physiologist called Arthur Galston worked on plant hormones and growth hormones. These are very interesting because they're important to understand if you want to know how plants grow, and they have potentially enormous uses in agriculture. Galston became incredibly disillusioned in the late 1960s when quite unexpectedly from his point of view some of his own research was used to create Agent Orange, a defoliating agent used in the Vietnam War. Here's a scientist saying I thought I was doing something that would have nothing but good use. Somebody has turned it into a dreadful weapon of ecological destruction. That, I think, indicates some of the dilemmas that science has. It's not so easy to be sure that everything you do will be used for entirely benign purposes and this is a dilemma.

The Responsibility of Scientists

Let's not be too hard on the scientists. Francis Bacon, way back at the beginnings of modern science said something profoundly true. "Knowledge," he said, "is power." We also have a phrase in English that power tends to corrupt.

The question always has to be asked, "What are we using our knowledge for?" I think you have to ask that question at every stage. Since knowledge can have unpredictably powerful effects there's an obligation on us all to be engaged in a debate about how we're going to use our science; preferably for better purposes. I don't think we can just offload all the responsibility on the scientists because very often their job is done when they have published a piece of research showing that such and such a thing is true in the natural world. That then becomes a piece of public knowledge; somebody else can pick that up and do something else with it. We need a much wider sense of responsibility and engagement with science.

> In the brief interval between the end of the Second World War and the present, there has been a profound transformation of the public regard for science. Half a century ago, there was general optimism that research and developmentwould quickly solve long-standing social and economic problems—not only impending resource shortages in the industrialized world but the continuing impoverishment of most other countries; now there is a sense of public irritation that science is apparently incapable of providing the sharp prescriptions for the management of the world, even in such matters as global warming, that would comfort the now substantial army of global environmental diplomatists.
> —John Maddox, *UNESCO World Science Report 1998*, p. 18

My Hope for Future Debate

Science has always been a global enterprise in the modern period. We do need new kinds of debate and new kinds of forums for discussion of what we're going to do with our science and technology. The relationship between science and society in many parts of the world, on many issues is not comfortable today. Thank goodness we don't live any longer in a society where most people are prepared to defer to the experts to sort it all out.

I think UNESCO (the United Nations Education, Science, and Cultural Organization) and the ICSU (International Council for Science), the organizers of the World Conference on Science for the Twenty-first Century, are in a good place to facilitate those new kinds of debate. It's got to be a world debate; most of the issues are international or global, they can't be confined to a single country.

RITA R. COLWELL

The Role of the Citizen

Rita R. Colwell

Dr. Rita R. Colwell is the Director of the National Science Foundation (NSF), an independent federal governmental agency in the United States that provides support for research and education in science, mathematics, engineering, and technology.

Trained as a bacteriologist and geneticist, Dr. Colwell holds a Ph.D. in Marine Microbiology from the University of Washington. She was Professor of Microbiology at the University of Maryland and President of the University of Maryland Biotechnology Institute before taking office at NSF in 1998. She has been awarded twelve honorary degrees, has held several honorary professorships, and has authored or coauthored sixteen books and more than five hundred scientific publications. She also produced an award-winning film, *Invisible Seas*.

She has been a member of the National Science Board from 1984 to 1990 and has held numerous other advisory positions in the U.S. Government, as well as in the international community.

In this brief conversation, we look at the fascinating world of micro-organisms and their many contributions. We are reminded of the enormous contributions science in general has made to the well-being of humankind. If the public is to understand this great impact, scientists and engineers must help to communicate the excitement of this work to society.

Dr. Colwell, how does someone become a microbiologist, and why? Give us some insight into the motivations and the challenges of microbiology?

I started out to be a chemist, but I found that the subject was not as exciting to me as some other areas. During my undergraduate training, I took some life science courses. The most exciting to me was a bacteriology course.

The world of micro-organisms discovered by looking through a microscope is absolutely fascinating. These tiny little creatures dart about actively in their drop of water. This vision through the microscope was probably a seminal moment, if you will, when I realized that life as a microbiologist was going to be fun—understanding bacteria, their genetics, how they change their environment and how they are changed by their environment.

Invisible Seas

Micro-organisms are an enormous unseen power. Some years ago I produced a film for my students called *Invisible Seas* about the bacteria that inhabit the oceans. Millions and millions of bacteria carry out change every day through their involvement in the cycles of carbon and nitrogen in the atmosphere and by producing compounds that influence the formation of raindrops and clouds. Micro-organisms affect the fertility of soil by fixing nitrogen from the atmosphere and bringing it into the soil to be used by the plants. That's the enormous unseen power of micro-organisms that inhabit the Earth.

It's so important to convey to the public that the micro-organisms that make us sick are only a very narrow proportion of the total population of micro-organisms. By far most of them are very busy doing good things for us. To eradicate all micro-organisms would be to essentially cease life on this planet.

Micro-organisms and insects would certainly survive a nuclear holocaust and they would recolonize very quickly. That's one of the exciting things about the new field of astrobiology, determining whether there is life on Mars, whether there are living creatures on other planetary bodies. Surely there must be. But the question is, what kind of life? Most likely, they're the sort of micro-organisms we find in Antarctica or in the Arctic regions at the poles. It is less likely that they will be the micro-organisms we find very happily growing in Tahiti.

The Role of the Citizen vis-à-vis Scientific Research

It's absolutely imperative in a democratic society that people are fully informed and understand science, engineering, and technology and their tremendous capacity. It's important that people are able to connect basic science and engineering with the availability of cellular phones, computers, and telephones, with access to food that's nutritious, and with access to medicines that combat infections that a hundred years ago were devastating. All of these advances provide a standard of living that's unparalleled in history. They are all based on advances and discoveries in science, engineering, and technology. These things didn't just happen. They're the results of the long hours of hard work and the exciting moments of discovery by scientists and engineers in understanding nature, explaining and utilizing it in a very positive way for the improvement of the wellbeing of humanity.

The Public's View of Science

Science has really contributed enormously to the wellbeing of humanity. Think of aeronautics, for example. A hundred years ago it would have been considered absurd to think that we would soon be flying in the air. You would also be considered rather strange to envision such conveyances as our private automobiles, built to provide so much comfort while traveling great distances. The advances in medicine are extraordinary. Look at our ability to understand the genetic basis of disease and to move into an entirely new area of genetic medicine. The benefits have been huge.

The negative aspects of application are, in comparison, very small. The problem we have to overcome is to make it clearly understood that the benefits are enormous considering whatever adverse effects that may have occurred and to work to minimize those effects, such as overuse or misuse of chemicals or antibiotics. That isn't to simply gloss over the applications, but in any human endeavor, there are those who will misuse and those who will use abundantly and wisely for the betterment of humanity.

Science, engineering, and technology aren't really understood by the public. On the other hand, we know from polls taken through funding provided by the National Science Foundation that at least 75 percent of the public supports basic research. They don't understand it but support it because the public understands that basic research ultimately will lead to improvement of the human condition as it has done in the past.

The Need for Civic Scientists

A decade or more ago scientists and engineers did not believe it was their job or their responsibility to explain what they were doing in their laboratory. That has changed.

There are more civic scientists, as my predecessor Neal Lane calls them, who spend time in the community to explain what their research is about, where it's leading, and the benefits accruing from it. This is a responsibility for all of us, as scientists and engineers, to make it clear that laser surgery for cataract removal draws from the fundamental research done in physics laboratories 20, 40, or even 50 years ago. Medical imaging that allows us very precisely to determine where a tumor is located in the brain, again, comes from chemistry, computational science, physics, and mathematics. The public hasn't made that connection. It is an enormous responsibility and a duty for scientists to communicate this to the public.

> Although scientist and nonscientist alike can marvel at the power of our knowledge in science and technology, it is the intersection of this knowledge with the goals and needs of society that is our larger responsibility. Understanding this crossroads of knowledge and needs and then acting on behalf of society will present our most challenging task.
>
> In the past few years I have spoken to many groups of my colleague scientists and engineers about a new, additional role that, I believe, we must play in society. I termed this role the "civic scientist," with civic meaning "concerning or affecting the community or the people." In this new civic capacity, scientists and engineers step beyond their campuses, laboratories, and institutes and into the center of their communities to engage in active dialogue with their fellow citizens.
>
> —Neal Lane, *AAAS Yearbook 1999*, chapter 22

Scientific Esperanto

I've conducted research to understand the epidemiology of cholera, a disease that is massive and global in scope. This research was quite interdisciplinary, involving oceanography, epidemiology, ecology, and clinical medicine, among others, including microbiology. I have found that meteorologists, physicians, and microbiologists talk to each other, but they use a language of their own discipline that is not really understood by all of the disciplines. They might as well be speaking German, Spanish, or Chinese.

We need to develop what I call a scientific Esperanto, a common language among scientists and engineers. That language has to be one that's also understood by the public. That language is explaining the research that's done, its benefits, its potential, and how it is done. To understand the scientific method is critical, the need for controls, the need for experimentation, the need for repetition of experiments, and the need for confidence of the results before they are put into practice. Unfortunately, the television depiction of scientists is that the discovery is made quickly and effortlessly in the laboratory, and should be immediately applicable, without further testing for safety, and so forth.

Prospects for the Twenty-first Century

Which takes me back to my own discipline. The role of micro-organisms at the molecular, genetic, and genomic levels has proven to be entirely fascinating. I think a fantastic area of research is opened up by understanding the complete genomic sequence of all those micro-organisms that comprise the communities that work so hard in this hidden, unseen world and do so much work to make this a sustainable planet. Understanding how all work together and their genetic basis is a wonderful prospect.

The patenting of a bacterium by Ananda Chakrabarty, who was working at General Electric at the time, set a very important precedent. We had previously patented strains of corn and varieties of tomatoes and petunias. It seems quite feasible that we patent those micro-organisms that carry out fundamental metabolic steps as we have done for the production of penicillin and their analogues. This has benefited research because it has induced investment in research for the next stage of drug therapy or for the next stage of genetic therapy that is so very important.

The Endless Frontier of Science

The beauty, the joy, and the frustration of science is that it is an endless frontier because it builds on information that is gathered iteratively. Then, more experiments are done and more information is gathered. You have another opportunity to pull from the accumulated and integrated knowledge, a basic principle, a fundamental concept, but that only gives you ideas for the next iteration. Science is an endless frontier, and I do believe that the human condition can be so described as well.

References

Lane, Neal. 1999. "The Civic Scientist and Public Policy," in *AAAS Science and Technology Policy Yearbook 1999*. Washington D.C.: AAAS. Also: www.aaas.org/spp/yearbook/chap22.htm

LEON M. LEDERMAN

Particles and Poverty

Leon M. Lederman

Recognized as one of the world's foremost experimental physicists, for several decades Dr. Leon M. Lederman has designed and performed experiments that have led to major breakthroughs in our understanding of the basic structure of matter. In 1962 Dr. Lederman designed and performed the experiments that earned him a shared Nobel Prize (1988) with Melvin Schwartz and Jack Steinberger for the discovery of the muon neutrino. Under his direction a long-lived neutral K-particle (1956), the first antimatter compound (1965) and the bottom quark (1977) were discovered. As Director of Fermi Lab (1979–1989) he supervised the construction and utilization of the first superconducting synchrotron. Since his retirement he devoted his considerable energy and enthusiasm to education, setting up educational centers for science teachers and gifted students.

Scientists do science because the universe is revealed as beautiful and mankind is a part of this beauty. We are also aware that only through science can we create the utopian Garden of Eden for the planets' inhabitants. Finally, we are very concerned about the dark sides of science and technology, about environment, population, superstition and rigid belief systems, the growing gap between technological have's and have-not's, the alienation and fear of too much-too fast—all of which speak to the urgent need for universal science literacy.

Professor Lederman, take us back—long before your Nobel Prize—to the beginning. What sparked the young Lederman's interest in understanding the very structure of the universe? And what is important to you today?

I was interested in science as a child. I remember a book written by Albert Einstein that was probably for children in which he compared science to a detective story where there's a mysterious set of clues, a bloody glove, a white car, or a barking dog, a set of mysterious circumstances out of which the scientist or the detective finds a rational explanation. Science is a search for rational, systematic, logical explanations for phenomena. That notion of science intrigued me.

My choice for physics came only gradually. I was influenced by fellow students who were interested in science and I said, how come you guys are interested in science? They were interesting and I wanted to hang around them. A strong influence in high school was not a teacher, but a young guy who had a job fixing up the laboratory. He was working on his Ph.D. thesis at night. He was much younger than the average teacher and that was very meaningful for us. We used to hang around him and he told us stories about his own research, which was inspiring. So I found myself in college majoring in chemistry because of that experience.

Slowly there emerged a whole new group of friends, some of them outstandingly brilliant and interested in physics. I thought the physics kids of my age were more fun than the chemistry kids. They were livelier, they were interested in other themes, music, and so on. Chemistry was very complicated; organic molecules were too complicated. Physics was very simple. Can we understand one particle and how it behaves and why it behaves the way it does? That seemed to me a more tractable subject, something I could really understand.

I finished college majoring in physics. Then I had to go help Eisenhower win World War II. Coming back, I remember being very lucky on the troop ship home, I won a lot of money. When the ship landed I was able to hire a taxi that took me to Columbia University in New York, and I registered for graduate school in physics and the rest is history, as they say.

My Nobel Prize was for a discovery that had to do with subnuclear particles called neutrinos, a name given by Enrico Fermi. The puzzle was very prominent in 1959–1960, and we solved the problem. We opened up a whole new field of exploration with neutrinos.

High-Energy Physics

High-energy physics is the study of the fundamental structure of the physical universe. It's called high energy because those are the tools we use, whereas astronomers will use telescopes and nowadays space observatories. We use particle accelerators. We smash nuclei as hard as we can. The energy is essentially the force with which two particles will smash together. It's the only way we have of seeing what is inside. A particle accelerator is like a giant microscope working not with light but with protons. The higher the energy of the protons the smaller the distances you can see. So there's a kind of trade-off. If I want to see extremely small distances, I have to use huge particle accelerators that cost huge sums of money and therein lies the difficulty.

Quantum Theory: The Single Most Poignant Advance of Knowledge

Democritus proposed the existence of atoms twenty-five hundred years ago and everybody loved atoms. Isaac Newton loved atoms. Galileo loved atoms. Ever since, through the whole Renaissance and beyond, atoms were considered a good idea, but nobody could come to terms experimentally with them. But around 1900, plus or minus a few years, we began to get data from the atom and the data was absolutely bizarre. It was more spectacular than the Starship Enterprise on Star Trek, landing on a crazy planet where apples fall up and things never work. The enormous intellectual armaments given to us by Newton and Maxwell that was so successful for hundreds of years didn't work inside the atom. It became such a great crisis that Wolfgang Pauli, the enfant terrible of modern physics of his time, said he should have become a radio comedian. He felt that physics is terrible because nothing is understandable.

It took twenty years of the best minds in Europe, Max Planck, Louis De Broglie, Erwin Schroedinger, Werner Heisenberg, Albert Einstein, and many others, to solve this problem of quantum theory. Suddenly a whole new perception of nature emerged by 1930, the end of the quantum revolution. And out of that came a total understanding of the atom and therefore a total control of the atom.

Fundamental Research and the Alleviation of Poverty

If a congressperson pesters me with why I'm spending so much money, I will simply say, our understanding of quantum theory accounts for some huge fraction of the gross national product of all industrial powers. In any portfolio of investments there must be long shots. There must be some fraction of your investments that have long-term possibilities. The fundamental research we do has to be looked at as a long-term investment.

When the electron was discovered in 1897 as a curious object that seemed to be connected to the structure of matter, the English professors who celebrated this discovery in Cambridge used to have a toast. They'd say, gentlemen, we drink to the electron. May it forever remain useless! Yet, the electron has been the most used beast of burden in our technology. It opened the door to all of technology. You can't give up on a quest for fundamental knowledge. But, we have to balance it. We have to have compassion for people who are hungry, but we have to have an investment that might solve this problem in unexpected ways.

The Coming Energy Crisis

Despite all advances there are still two billion people on the planet who live without electricity. Suppose we could somehow raise the quality of life for them to some reasonably higher level. The energy consumption per person would burden our environment in terrible ways and we don't know how to solve that problem. We do not have a solution for ten billion people, all living at a level where there is some pleasure in their life. Whether that answer will come from what we're doing in our search for the properties of quarks or not, I don't know, probably not. But, who knows?

The quest for fundamental knowledge has to be part of our portfolio in balance. We spend a lot of money on more obvious means for understanding the energy crisis. Some fraction of our expenditures to add to our knowledge base must be to understand the universe in its most thorough way. It's that motivation that has given us everything we have: television sets, medical devices, the fact that we can see at night, and that people of very modest means can push buttons and listen to Mozart or Shakespeare. It's these enormously fulfilling activities and the increase in longevity of thirty, forty years for all people who are lucky enough to have access to modern medicine.

The Blessings of Modern Science and Technology

We can look at quantum theory as the major breakthrough of the twentieth century, along with relativity. Those two theories are now the pillars of our understanding of the universe. We could look at our entire technological society as being dominated by the usefulness of these discoveries for making all kinds of conveniences. Medical devices that allow us to look for tiny tumors, the CAT scanners. The whole subject of medical instrumentation has been revolutionized.

The Internet is one of the things that came out of the particle laboratory at CERN in Geneva where they invented the World Wide Web. That revolutionized the way we in which behave, the way we think, and the way we talk to each other; it's an enormous device for having information access and communications between people and so on. All of the things we take for granted around our lives, communications, and the ability to listen to great music by the touch of a finger, the ability for modest people to travel the world and go to various exotic places. The universality of education—millions and millions of people now expect to go to college whereas, fifty years ago, it was only thousands.

> The dream behind the Web is of a common information space in which we communicate by sharing information. Its universality is essential: the fact that a hypertext link can point to anything, be it personal, local or global, be it draft or highly polished. There was a second part of the dream, too, dependent on the Web being so generally used that it became a realistic mirror (or in fact the primary embodiment) of the ways in which we work and play and socialize. That was that once the state of our interactions was on-line, we could then use computers to help us analyse it, make sense of what we are doing, where we individually fit in, and how we can better work together.
> —Tim Berners Lee, inventor of the World Wide Web, www.w3.org/People/
> Berners-Lee/ShortHistory.html.

The Shadow Side of Modern Science and Technology

There are however many dark sides to these blessings. Some of them could have been anticipated had we been wiser, and others were very difficult to anticipate. The ecological damages that we induce by our industrialization, and by our carelessness in the emission of wastes. We've given people this gift of transportation that is amazing. Relatively poor people in the developed world can travel vast distances freely in their cars. The car was originally designed as a non-polluting device to replace the horse. But now, the pollution from cars is endangering the future of humanity on this planet because global climate change is a very serious issue.

Technology is exacerbating the gap between rich and poor because if you are rich, you have access to technology and if you are poor, you don't have it. The more access you have, the richer you become. The gap between rich and poor nations and between the rich and the poor in a nation is getting worse. The population explosion itself is a result of our science and technology. Immunizations and sanitation, those relatively simple things have created a world population of six billion people.

Nuclear weapons are another example. I'm a member of the board of the Bulletin of Atomic Scientists that has the doomsday clock that we just moved closer to midnight because of the tests in East Asia. We've been relaxed a little bit because the Cold War is over, but we still have tens of thousands of nuclear weapons and that's a menace to humanity. You can argue and defend the fact that we built them, but you can never defend why we built so many.

> The Doomsday Clock first appeared on the cover of the June 1947 issue of the Bulletin of the Atomic Scientists. It was set at seven minutes to midnight—a decision that the Clock's designer, Martyl Langsdorf, said had more to do with principles of good design than with the current state of the world (the very idea of the last quadrant of a clockface was supposed to represent in itself the urgency of the situation). Since that first cover, the hands of the Doomsday Clock have been moved 16 times. The last time was in May 1998, when India and Pakistan each test a series of nuclear devices. The clock move showed that deep reductions in the numbers of nuclear weapons, which seemed possible at the start of the decade, have not been realized.
> — The Bulletin of Atomic Scientists web site, www.bullatomsi.org

I think it's fair to say that almost anything technological is a two-edged sword, which benefits humanity and is calamitous or disastrous at the same time. Who makes the choice about how to use technology? It seems to me that one of the encouraging things about our world is the growth of democracy. We have some hope that we're leading to a worldwide appreciation of democratic societies and if we have democratic societies, then we know who makes the choice. It's the people who make the choice.

Science Literacy

At the moment, it isn't the people who make the choices because the people are not informed enough. Here we get to my favorite subject of universal science literacy. If we're going to survive in this complex world of enormous technological capabilities, we have to have a science literate general public. I have confidence that if the public is educated, they will win out and good things will happen. The only chance we have is to educate the population. It's a lot easier than to educate the politicians.

Educational Outreach

Our largest project is with the Chicago public schools. There's a Third World arena if you like, embedded in an extremely rich city side by side with the monuments, banks, and corporations. There are the ghettos and the slums where extremely poor people live in dangerous streets with crumbling school buildings. We know that education is the lifeline out of this trap that they have of dropping out of school, unemployment, crime, drugs, and teenage pregnancy. A deadly cycle that you can't get out of except through education.

Education works beautifully when you start with children. All children are scientists because the scientist is judged by the questions that the scientist asks. Answers will come later, but the questions are crucial. A good scientist asks good questions and kids ask good questions. It's that curiosity that the child brings to the classroom that's the most precious thing of all. They try to understand why a ball bounces, what drops, and what doesn't bounce. How strong are materials? How can you break them? Children don't have to do chemistry and biology; that's done automatically, but through physics they're learning about the world so it is a very practical subject.

> Science teaching in schools is expected to play a key role in the development of knowledge and in the general public's understanding of science. This could usefully be accompanied by awareness-raising initiatives and informal training and education. For some time "science weeks" or "scientific and technological culture week" have been organised in France, Italy, The Netherlands and Sweden. Similar operations have been carried out more recently in Portugal and Germany. In the UK a science week based on the same model has been added to the Edinburgh festival and to the annual festival of the British Association for the Advancement of Science. Since 1993, a "European week for scientific and technological culture" has been staged each year at the initiative of the European Commission.
> —*UNESCO World Science Report 1998*, p. 78

I try very hard to make my physics relevant to these children. For example, the greatest hero in Chicago is Michael Jordan, probably the greatest hero in the world. And when he jumps up in the air he seems to hang there in the air for a while. It's called the hang time. That's a good calculation. With a little bit of algebra, the beginning high school kid can calculate how long Jordan stays up in the air.

This is part of science, the practical things around you, things that kids are appreciative of. But one should never neglect the basic romantic notion of the world that children bring to the classroom and one should not disappoint them. They want to know about the stars and why the moon is full sometimes and half other times.

More and more we are training teachers to say to the child, that's a good question. I don't know the answer to that, but I know how to find out. And they can use the Internet.

My Hope for the Future

Physicists are genetically distorted with this optimism gene. I think that the world will, on balance, be more and more sensitive to the dark side of technology. Even the corporations that are always accused of excessive greed and anxiety for profits are becoming more public spirited. They become more cognizant of the fact that in the long run they're better off if we take care of our environment, clean the water, and the air, and lessen the bad effects of automobile exhausts and some of the many other things they do. That kind of responsible corporate leadership is growing.

People are gradually becoming more sophisticated and less interested in things that worry me a lot that is what we call junk science: fortune-tellers; witchcraft; some of the crazy alternative medicines such as peach pits to cure cancer; the charlatans that prey on the ignorance of people. Education is a key, because one of the key things when we teach kids science is that we teach skepticism, looking at the data critically. What's your proof? Where do you get the data?

Half of the learning of science among children should not be the content of science but the process of science. How does it work? Why is it so successful? What are the ingredients in it? And there you get skepticism, estimation, and making predictions. What will happen if we do the following? This is what citizens have to know to take care of themselves, to take care of the community, and to be effective citizens in a nation that is on the frontier of technology.

References

Berners-Lee, Tim, with Mark Fischetti. 1999. *Weaving the Web: The Original Design and the Ultimate Destiny of the World Wide Web by Its Inventor.* New York: HarperSanFrancisco.

The Bulletin of Atomic Scientists, www.bullatomsci.org

The Leon M. Lederman Science Education Center at Fermilab, www-ed.fnal.gov/ed_lsc.html

JULIA MARTON-LEFÈVRE

Only People Give Me Hope

Julia Marton-Lefèvre

Julia Marton-Lefèvre was born in Hungary and educated in France and in the United States, receiving degrees in History and Ecology and in Environmental Policy. Marton-Lefèvre is currently Executive Director of LEAD (Leadership for Environment and Development) since 1997. This is an international organization that provides outstanding men and women in mid-career with the training designed to enhance their leadership capabilities and knowledge of issues related to environment and development. Prior to this position she was Executive Director of the International Council for Science (ICSU) in France.

She serves on a number of international boards: the China Council for International Cooperation in Environment and Development, the Board of the World Resources Institute (WRI) as Vice-Chair, the Corporate Environmental Advisory Council of Dow Chemicals, and the ICSU Committee on Science and Technology for Development (COSTED). She served on the United Nations Secretary General's Task Force on Environment and Human Settlements and has been involved in a number of international bodies concerned with environment and development issues, such as Earthwatch, the Earth Council, the Center for Our Common Future, and the World Conservation Union (IUCN). Before joining ICSU, she was a Program Specialist at UNESCO in Paris working on issues related to environmental education, and prior to that she was a Peace Corps teacher in Thailand.

In 1999 she was the recipient of the American Association for the Advancement of Science (AAAS) Award for International Cooperation in Science where this interview took place. In this interview, she describes her experiences running the world's oldest and most distinguished organization of scientists (the ICSU) and now a younger body (LEAD) devoted to identifying, training, and networking future leaders from all corners of the globe. Clearly networking has been a key in her experience, as has a strong belief that people, when motivated, informed, and working together, can make a difference.

Ms. Marton-Lefèvre, you have devoted a considerable part of your professional life to international scientific dialogue and collaborations; to the very essence of cooperation. Give us some insights into the problems and promise of such collaboration.

Science by its nature is international. That is not always so obvious to some great powerful countries like the United States, but it is international. Scientific cooperation has been going on ever since scientists started to do science. The formal cooperation among scientists through ICSU, the International Council for Science, is in fact a hundred years old this year (1999). Very few people know that. So I think the basis of that cooperation is already there.

What governments have to recognize is that they have to provide more facilities for scientists to work together; they have to provide more funding and open doors to help achieve this. We from the scientific community have to be a bit more organized also, I think on two levels.

First we have to increase our interdisciplinary communication. Even today there are more scientists comfortable working in their disciplines speaking the language of biology, physics, geology, and geography. There are a number of large interdisciplinary programs that are now well on their way where that is happening. Second, we have to go beyond the natural sciences so that a bridge is established between the natural and the human sciences, the social sciences. That has not made as much progress as the link between the different disciplines within the natural sciences. I think that will be a very large challenge.

Finally, we have to make sure that younger people get involved in these issues and that they're trained as scientists or that they are trained as decision makers who are interested in science and who will allow scientists to work together.

What I Am Most Concerned About

We need to work together, to address the global issues of environment and development. The two cannot be separated anymore. Formally governments in 1992 accepted that, but they are still treating them as different departments. We need to work together to prove to the policymakers, who are the people who finally give the money for all this, that the problems require large-scale solutions and a large-scale involvement. We need a facility for the biologist to work with the physicist, to work with the social scientist, and to work with people from the media so that people are sensitized.

We now have quite a lot of scientific understanding and underpinning of what is wrong with our planet and the environment and development areas. What we now need is to make sure that we can communicate these findings so that people start behaving differently. Not everyone needs a car, not every family needs five cars, as happens in some countries. We need to mobilize all of society to communicate this message so that human behavior is slowly and steadily changing to correspond to the findings that scientists are proving.

One of our big problems is of course that not everyone believes the findings of science. We the scientific community have very prudently said, we're not sure, so let's take the precautionary way. The precautionary principle means that if we're not 100 percent sure about our findings, but we suspect their validity strongly, for example, that there is a global warming because of human intervention, we say, let's be precautionary. Let's start to change our behavior so that when the scientific proof is absolutely there, human beings are already prepared and the appropriate action has been undertaken.

So you can see why scientists cannot work on these problems alone. They need people to help communicate those ideas; they need social scientists to start working on changing human behavior. And we need the policymakers to ensure that the appropriate decisions are made.

In many cases, NGOs have proved more adept than both governments and the free market at responding to human security needs. In Bangladesh, where 5,000 NGOs are involved with literacy programs, a child is more likely to learn to read with the assistance of an NGO than through a state organization. Worldwide, NGOs now deliver more development assistance than the entire UN system.

—Curtis Runyan, *Vital Signs 1999*, p. 145

Communicating Science to the Citizen

It is of course a major problem for the scientific community, to bring all that together and to inform citizens of scientific findings. In some parts of the world where the economy is going very well, the price of oil for your car is very inexpensive, so people don't understand the bother about environmental concerns. They easily consider buying another car, driving another five hundred kilometers or miles. But in some parts of the world where the economy that drives the whole system is going through a difficult phase, the scientific message is able to get through better.

Unfortunately television powerfully influences consumer habits. Seeing how people live in some parts of the world makes it look so easy. I know scientists all over the world and many of them, especially in countries like India, for example, are actually involved in activities beyond science. They appear on television, doing science education, talking about environmental problems and the behavior that needs to go along with solving them. To me it seems that they get more airtime and newspaper articles than they do in the United States, for example.

I was in China recently and every day there are dozens of articles about environmental problems in the various English language newspapers. I am told that this is also the case with Chinese television and the Chinese press. So scientists in those countries have been able to pressure the media and the "decision makers" to pay attention to this, so that they can communicate to the community that it is a very short-sighted solution to live like the people they see on television.

Regarding the response by government leaders and the public, I must say I am not a pessimist. My worry is aimed at our own community itself. Scientists are often very busy trying to find those answers. They don't see that it is necessary for them to come out of their laboratory. My feeling is that, although the scientific community hasn't done too badly, it needs to organize itself even better to make sure that the message leaves the laboratory and goes beyond the scientific community to the right target audiences.

The Yearbook of International Organizations documented 985 active "international NGOs" in 1956. By 1966 that number had increased to more than 20,000.

Half of all NGOs in Europe have been founded in the past decade. The number of NGOs in the United States is now estimated at 2 million.

—Curtis Runyan, *Vital Signs 1999*, p. 145

Science and Gender

I wouldn't say there is a gender *bias,* but there is a gender *problem.* The problem is the problem of ourselves, the women. Frankly it's a very full-time all-consuming activity to be a scientist or even to work in any profession. I have managed somehow having a family and having children. I have only two children; I don't think I could have done it with three or four; but it takes a tremendous amount of effort on the part of women to be there. Once we're there I don't feel the bias at all. I was the very first woman to direct ICSU, not because there was a bias but probably no one else thought they could do it. I wasn't hired because I was a woman. But it also wasn't a handicap.

How I Got Involved in Science

I'm not a real scientist, I'm really a science policy person. I started by studying biology and history together. I always felt that history is very important. I come from a very small country; my family moved around and I needed to understand how we got there. I was interested to find out how human beings function within their historic society and that was biology.

I was interested in how we evolved to be members of our society, which then determined the way in which history evolved. That combination was an unusual degree that would not have got me anywhere. This was a time when much older people were preparing for the Stockholm Conference on the environment. I, like many other college students in those days, read Donella Meadows's *The Limits to Growth,* and Barbara Ward's *Only One Earth,* and they really inspired me. That is when I started to study the environment.

Of course the environment wasn't considered a discipline. When I first made the shortlist for Assistant Executive Secretary of ICSU, the International Council for Science (of which I became the Director later on), I was interviewed and my scientific discipline was environmental planning by that time. The very distinguished Secretary-General of ICSU, a Nobel Prize winner and a wonderful man, was a little suspicious. He said, what is this? This is not really a science! I was selected in spite of this.

Later on as ICSU started to do much more in science, and started to launch the greatest global programs on global climate change and the environment, obviously my background was very helpful. In fact my background was appropriate because it gave me a chance to see the international scientific community in its totality, much like the way we look at ecosystems.

The International Scientific Community as an Ecosystem

Probably my academic training and also the kind of person I evolved to be through my genetic evolution and being a multicultural person, gave me the tools to look at all of the elements, or at least as many as I possibly could of that international ecosystem; an ecosystem that was made up of disciplines and cultures that were different and couldn't talk to each other. I think I had the gift of being able to listen to the different noises and voices and to try to come up with a program that everyone would be willing to agree upon. I had the skills to put together a good program to which contributions came from all the different parts of that ecosystem.

A New Generation of Leaders

Can I still sleep at night, seeing all that happens to the planet? I must tell you a little bit about my new job. My job is concerned about identifying talented young people who we know must be out there. In my previous job with a more established organization consisting of all of those distinguished scientists, the younger people were not involved. I was convinced that in order to find solutions dealing with future generations we had to find the successor generation of leaders of the future generation; not only leaders in science as I was convinced that the problems were more complex than just one community of science.

So we had to involve the media; we had to involve economists, businesspeople, NGO leaders, academics certainly, and government officials. My new job consists of identifying those leaders and that is difficult. It's something like an international headhunting organization. It's not so easy to find future potential agents of change in countries where we are not really used to working—Nigeria, Zambia, China, India, Indonesia, Pakistan, Mexico, and Brazil.

We eventually find those people. We give them a fellowship to study more, to open their minds beyond their disciplines, beyond their sectors and most of all beyond their own home communities. Media people usually know other media people; economists know other economists; scientists know other scientists. We bring these young people together. I'm not talking about kids, I'm talking about people between thirty and forty.

What makes me sleep better at night is that I know that these people exist. They must be given a chance to contribute to the way in which international programs are designed. They must be given a chance to be heard from, then their energy and talent, and their commitment to changing the way the planet is run is going to make a difference.

A New Style of Leadership

I see that this will bring on a new style of leadership that is sensitive, responsive, and articulate, and that cares about making needed changes. Indeed that is what allows me to sleep at night. I don't find articles about those kinds of things in the newspapers. If I would read newspapers or watch television I think I would be depressed. But I know that there are very, very good people out there. People who care and who are talented and energetic, and who are going to make a difference in their behavior.

We just need to make sure that more of those people are heard from and that more of those people are on the world scene. They are going to get to know each other through my program, for example, and through others like this. There will be a whole group of like-minded individuals who perhaps make a difference in our children's generation.

References

LEAD (Leadership for Environment and Development). web site: www.lead.org

III

Science and Peace

FRANS B. M. DE WAAL

The Biology of Making Peace

Frans B. M. de Waal

Professor Frans B. M de Waal was born in The Netherlands and received his Ph.D. in biology from the University of Utrecht in 1977. Professor de Waal began studying the behavior of primates. In 1981 he accepted a research position at the Wisconsin Regional Primate Research Center where he began studying reconciliation behavior in monkeys. He received the Los Angeles Times Book Award for *Peacemaking among Primates* (1989). His current interests include the evolutionary origin of human morality and justice. On this subject, his books raised much interest: *Good Natured: The Origins of Right and Wrong in Humans and Other Animals* (1996) and *Bonobo: The Forgotten Ape* (1997).

We have a tendency to look at aggression as antisocial behavior, as something separate from social relationships. But it is part and parcel of the closest friendships and the most cohesive families. Aggression is to be studied in this context: as the fire that keeps relationships moving, while at the same time having the potential to destroy. To prevent this, however, we have powerful ways of resolving conflict.

Professor de Waal, one of the hottest debates in the world concerns animal rights. You seem to have steered clear of the debate itself in order to examine one particular train of thought, namely, aggression and peacemaking. By spending so much of your life in the company of primates other than humans, what have you learned?

Konrad Lorenz wrote on aggression in the 1960s and right before him the science journalist Robert Ardrey had written *African Genesis*. Those books are very heavy on the aggressive side of human nature. After World War II that was probably a logical obsession for biologists. Both of them had a very pessimistic message: we are genetically aggressive and it's very hard to get rid of that. At the moment, however, there is a lot of new thinking of which one possibility is that much of the aggression that we see presently in the human species only came after the agricultural revolution. Whereas Lorenz and Ardrey were speculating, based on fossil finds of Australopithecus that have gone back much further.

Many people are not convinced any more that Australopithecus was necessarily a very aggressive creature. Raymond Dart said he was carnivorous, which is not the same as being aggressive and engaging in warfare. Eating prey is not the same as killing con-specifics. There is a lot of new thinking based on new fossil evidence that seems to contradict this notion that our history is one of bloodshed going back six million years, when we split off from the chimpanzee and the bonobo.

Hidden in the dense rainforests of Central Africa, Pan paniscus, the bonobo or pygmy chimpanzee, remained "undiscovered" by the outside world until 1929. It was then that scientists began examining a collection of skulls deposited in a Belgian museum. At first it was thought that these bones belonged to a new subspecies of the well-known chimpanzee, but taxonomists soon realized that they were dealing with an entirely new species of ape. In 1993, Harold J Coolidge proposed that it be classified as a full species, Pan paniscus: the fourth great ape.

Originally called the "pygmy chimpanzee", the preferred name is now Bonobo. The name's origin is unclear but it may be a garbled version of Bolobo, a town in Zaïre where scientists found the original specimens.

—World Wildlife Fund, www.panda.org

Aggression and Reconciliation in Primates

My work started in the period when Lorenz was very popular. As a graduate student I was given the task of observing aggression in primates. I have seen a lot of aggressive behavior so I don't have a rosy picture of primate behavior. I became interested in the opposite side because I observed that fights end. Life continues after conflict; friendships and family life go on despite the many battles. I became interested in how animals move from an aggressive state to a peaceful one; what kind of social and psychological mechanisms exist that make that transition possible.

What I discovered in chimpanzees was reconciliation behavior. Reconciliation means that two individuals who have a fight later come together, kiss and embrace, and then groom each other. They have a friendly reunion. That is a mechanism that we have now studied. I'm not the only one any more who is doing that. There must have been studies on twenty-five different primate species and we're looking for that even beyond the primates.

Other mechanisms are more difficult to get at, such as similar mechanisms as in humans. Forgiveness, for example, is a very popular topic within religion that is related to reconciliation. There is however an internal process that is very hard to put your finger on and not only in non-human primates, but we are currently involved in experiments that relate to forgiveness.

In the latter part of his career, Lorenz applied his ideas to the behavior of humans as members of a social species, an application with controversial philosophical and sociological implications. In a popular book, *On Aggression* (1963), he argued that fighting and warlike behavior in man have an inborn basis but can be environmentally modified by the proper understanding and provision for the basic instinctual needs of human beings. Fighting in lower animals has a positive survival function, he observed, such as the dispersion of competitors and the maintenance of territory. Warlike tendencies in humans may likewise be ritualized into socially useful behavior patterns.

—Encyclopedia Britannica, www.britannica.com

How I Got into Science

I've always liked animals so I wanted to work with animals. That's not to say I wanted to be a scientist, but it's a sort of motivation that you also find in people like Konrad Lorenz or Ed Wilson, or Jane Goodall. They have that sort of drive that declares: We love animals; we want to work with them. If, in fact, you have an inquisitive mind, that combination makes you want to become a scientist.

In the beginning, studying the chimpanzees of Gombe was not easy for Jane Goodall. The chimpanzees fled from her in fear, and it took many months for her to get close to them. With determination, she searched the forest every day, deliberately trying not to get too close to the chimpanzees. On many days Jane observed the chimpanzees through binoculars from a peak overlooking the forest. Gradually the chimpanzees became accustomed to her presence.

Jane uncovered many aspects of chimpanzee behavior during her first years at Gombe National Park. In October 1960, she observed a chimpanzee using and making tools to fish for termites. This discovery challenged the definition current at that time: Man the Toolmaker. Because of her research, we now know that chimpanzees hunt for meat and use tools. The longer the research continued, the more it became obvious how like us chimpanzees really are.
—The Jane Goodall Institute, 1986, www.janegoodall.org

As a graduate student I discovered reconciliation. No one had ever described this before. It was not a logical thing for a European ethologist. An ethologist is someone who studies animals from a biological perspective. We had all these theories that emotions have some sort of continuity. We looked at behavior as a string of things that are motivationally connected. Reconciliation is almost opposite to that and that was a bit of a discovery. It was out of the ordinary in the sense that you have an aggressive state, yet maybe two minutes after the fight, it turns around into an immensely friendly state. That was not logical from the traditional ethological perspective.

I used to say jokingly that the reason I'm interested in this field is that I'm from a family of six boys, so war and peace is a natural issue for me. I studied aggression but I've never looked at aggressive behavior in animals or in people as necessarily antithetical to social relationships. I've always looked at that as part of social relationships. Of course sometimes things get out of hand. But most of the time aggression is built into the relationship and functions within it, so I think this may be my background.

The Anthropodenial

There is still a debate going on about whether animals have intelligence, compassion, empathy, and so on. There are still quite a few who would categorize everything that even reeks of mind and feelings as anthropomorphism and who have a problem with that. I look at that argument as a rearguard debate of the behaviorists, of which there are still quite a few out there. I have taken a very strong position against this. At some point I spoke of "anthropodenial" as the opposite of anthropomorphism.

Take, for example, an animal like a chimpanzee. With an animal that is so close to us as this primate, the safe assumption is that if the chimpanzee does something very similar to us, then the motivations and the mental processes that go into that are very similar to ours as well.

For example, I have studied coalitions in chimpanzees and written a book called *Chimpanzee Politics*. Chimps are able to form coalitions to overthrow an established leader. You may say that to call that politics is anthropomorphism. I would argue that the mental processes that go into that must be very similar. That is a safe assumption because the species is so closely related; it diverged from us so shortly ago, that there was no time for totally independent evolution of such complex behavior.

Evolutionary Parsimony

There are two sorts of parsimony. The behaviorists are big on parsimony. They strive for cognitive parsimony, which means that you look for the simplest possible explanation. That is why they exclude certain mentalistic explanations. Behaviorists sometimes forget that disproving intentions, emotions, and thinking in animals is just as hard as proving it.

I would argue that you also have evolutionary parsimony, which says that if similar animals with a similar evolutionary background show similar behavior, you need to come up with similar models to explain it. This would mean that for everything that we do that is similar to what chimps do, we try to find some shared theoretical ground. From chimpanzees you can expand that to monkeys and it's not so difficult to extend it to other species, maybe dolphins and elephants, other large-brained mammals.

I think that the danger with regards to those species is anthropodenial. Due to an excessive fear of anthropomorphism, you don't want to use human concepts and so you're going to miss out on something substantially similar between what we do and what they do. It's a very old tradition in the West; it doesn't exist at all in the East. Anthropomorphism is not, and has never been, an issue for Japanese scientists, just as evolution has never been an issue for them because in Buddhism and in Eastern thinking there is a spiritual connection between humans and animals. Your soul can show up in some other animal at some point and can come back to a human. Anthropomorphism is a typical Western problem about which I have already said fifteen years ago I don't want to deal with it anymore. I think it's totally fruitless.

Conservation and Animal Rights

Even if elephants have no feelings, does that mean that they could be eliminated by us, that we could just shoot them and get rid of them? I think this whole issue of conservation and animal rights is not necessarily tied to these things. Some people do that, saying that if apes have language, they have rights. I think there are some serious concerns that apes, even the best trained apes in language programs do not possess something resembling human language. I would still argue that whichever way that issue is settled, for me it doesn't necessarily settle the issue about animal rights or how we should treat animals. We can respect animals and protect them or at least not hurt them, regardless of what certain psychological studies show.

A much more important issue is whether they can suffer. I feel like almost everyone else who works behaviorally with animals that they can. That's a relevant issue. I would argue in addition that you also protect or are concerned about animals or people simply for what they are.

Naming of Animals

With the resistance against anthropomorphism there has been a bias against the naming of animals as well. There was a debate about Jane Goodall doing that, and Albert Schweitzer was often called a sentimentalist. I think the naming of animals started in Japan, by Kinji Imanishi in the early 1950s. He was a founder of Japanese primatology and was interested in individuals in society and he started to give animals names. I have just come back from Japan where I was doing research for a book on the history of all this. People just didn't believe that he was able to recognize one hundred monkeys on an island. That was not conceived as possible. He had to convince them that it could be done. Now we know that it can be done because so many of us do the same thing.

Conflict Resolution

Many of us are embarrassed by conflicts that maybe two hundred years ago we would consider perfectly normal for the human species. After World War II, for example, we thought we could get over that and create a more peaceful world. However, at some places in Europe things started falling apart. I'm a European and I feel sort of embarrassed that this happened. That, for example, we didn't have the maturity to overcome a certain ethnic xenophobia that exists there and that will probably always remain.

In studies of non-human primates we have a basic model that I think works most of the time. There's some experimental and observational evidence that says that peacemaking tendencies are strongest between individuals who have a valuable relationship. That is called the valuable relationship hypothesis. For example, you can bring monkeys into a situation where they can only eat if their friend is with them. They can only get good food if they coordinate activities and go together to the feeder and so on. If you create that sort of situation between monkeys, you get a much higher reconciliation rate when you induce fights between these two monkeys. By creating a valuable relationship, you increase the tendencies for peacemaking.

Another factor that is important is what we call the "reparability of relationships." That's why, for example, people more easily get into fights with their spouses because those relationships are reparable. With colleagues relationships are much harder to repair. Those are factors that play a major role in the initiation of aggression and the making of peace.

An Example in Politics

What I see in, for example, the European Community is that they have basically taken the principle of the valuable relationship hypothesis and applied it to a number of countries. After World War II France, Germany, and the Benelux countries (Belgium, The Netherlands, and Luxembourg) considered the fact that if they made economic ties, perhaps they could reduce the possibility of aggression between their countries. If France were to invade Germany now, or the other way around, they would have a lot to lose. That is, of course, a possibility worldwide. There is this saying that if we have a common enemy out there in extraterrestrial space, then probably we would suddenly reduce the warfare on this planet. Most likely that's never going to happen but it is possible that by having a global economy where all these connections start to exist and valuable relationships are created globally, that you are going to reduce warfare.

The End of All Conflict?

Perpetual peace can never exist. It's a bit like creating a car engine without friction; that also will never happen. There's always going to be friction within systems. I don't believe that you can eliminate tensions and conflicts completely. You can channel them in particular directions, you can control them in particular ways, you can set up mechanisms, and you can set up relationships that temper the intensity of conflict.

Conflict is always going to exist. It's almost like someone who gets married and believes that marital life will be 100 percent peaceful. Most of us who are married will tell this person that that's never going to happen; there is no 100 percent peaceful marriage. And if it exists, it's probably a very boring one. I don't believe in systems without friction and ideas of a perpetual peace don't really appeal to me.

Are People Essentially Peaceful or Aggression-Driven?

The question of whether we are inherently aggressive or not is a Lorenzian sort of phrasing, which I would never indulge in. The question is, will we have a world without conflicts of interests? I would say we will not. Do conflicts of interest need to translate in aggression? That's where we can have a debate, but the debate previously was always phrased in terms of aggression as a sort of drive model.

This was a Freudian idea that Lorenz borrowed, aggression as a drive that needs to come out whatever the circumstances. It builds up and it comes out. The term "aggression" acquired an exalted aura as if it were something out of our control. I have a much more Darwinian view, where each individual, nation, or group has its own interests; interests that collide with another parties' interests so that there will be many ways of working things out.

I studied food sharing in primates. There are primates who share their food instead of fighting over it. They don't always do that, so when do they do that? It's a model with many options. You have conflicts of interests and it can go this way and then you have warfare. Or it can go another way, like using economic ties. There are many ways of resolving issues between individuals.

Research and Policy Making

For me, it's not important that my research affects policymakers. I do think, however, that there are two things that can be learned from this kind of research on primates. One is that we are building now more detailed theoretical models. We are testing them with experiments and observations and we may be able to apply those models to human children at some point. There are many studies now on children that use the same techniques.

The second contribution that we make concerns the techniques. We have observational techniques that are very detailed. People who do conflict resolution research on humans don't always apply those. They often use a sort of questionnaire method. For example, you ask a kid, Do you have a fight and what do you do with a fight? That's not the approach I would take. I would want to see what they do, because I don't believe, necessarily, what children or adults tell me.

References

Ardrey, Robert. 1961. *African Genesis.* New York: Dell.
Goodall, Jane. 1986. *The Chimpanzees of Gombe: Patterns of Behavior.* The Jane Goodall Institute. www.janegoodall.org
Lorenz, Konrad Z. 1963. *On Agression.* San Diego: Harcourt Brace.
de Waal, Frans. 1997. *Bonobo: The Forgotten Ape.* Berkeley: University of California Press.
———. 1982. *Chimpanzee Politics: Power & Sex among Apes.* New York: Harper & Row.
———. 1989. *Peacemaking among Primates.* Cambridge: Harvard University Press.
———. 1996. *Good Natured: The Origins of Right and Wrong in Humans and Other Animals.* Cambridge: Harvard University Press.
World Wildlife Fund, http://www.panda.org/resources/publications/species/greatapes/nathistoryC.htm

JOSEPH ROTBLAT

A Life Devoted to Peace

Joseph Rotblat

Born and educated in Poland, Sir Joseph Rotblat moved to the United Kingdom in 1939. He joined the team that worked on the atom bomb in Los Alamos until 1944, when he left the project as it became clear that Germany would not produce an atomic bomb. Sir Joseph Rotblat followed a distinguished career in nuclear physics and biophysics. He became famous as one of the founders of the Pugwash Conferences on Science and World Affairs, of which he was the Secretary-General from 1957 until 1973. He also edited numerous conferences of scientists and edited the Pugwash Newsletter.

Rotblat and the Pugwash Conferences shared the Nobel Peace prize in 1995, "for their efforts to diminish the part played by nuclear arms in international politics and, in the longer run, to eliminate such arms."

Excerpt from Sir Joseph Rotblat's speech at the World Conference on Science, Budapest:

Fellow scientists, should you be concerned about the ethical issues and the social impact of your work? Should you accept responsibility for the human and environmental consequences of scientific research? These questions did not arise in the distant past because there hardly were such consequences. In those days, science had no role in the day-to-day life of the people or in the security of states. Science was then largely the pursuit of gentlemen of leisure. They would collect plants or fossils; they would gaze at the sky and note unusual events. There was no Internet in those days and so they communicated their observations to other gentlemen with similar hobbies and gatherings had a social character, a sort of salon entertainment.

The impulse for those pursuits was sheer curiosity, the same that drives scientists today, but with no proclaimed practical ends. In the course of time, science began to be taken up as a full time profession. Learned societies and academies of science were established with highly exclusive memberships that widened ever further the detachment of scientists from society. This detachment let them to build an ivory tower in which they sheltered, pretending that they have nothing to do with human welfare. The aim of scientific research, they asserted, was to understand the laws of nature and since these laws are immutable and unaffected by human reactions and emotions, therefore, these reactions and emotions have no place in the study of nature.

Arising from this exclusivity, scientists evolved certain precepts about science to justify this separation from reality. These precepts included something like, science is neutral, science for it's own sake, science has nothing to do with politics, scientists are just technical workers, or science cannot be blamed for its misapplications. John Ziman in a paper on the basic principles of the social responsibility of scientists, a joint project within Pugwash and UNESCO, analyzed each of these postulates and found them all wanting in the current context.

The ivory tower mentality was perhaps tenable in the past when a scientific finding and its practical applications were well separated in time, the time interval between an academic discovery and a set of applications could be on the order of decades. And also it was implemented by different groups of scientists and engineers. Pure research was carried out in academic institutions, mainly universities. And the scientists employed there usually had tenure; they were not expected to be concerned about making money from their work. Taking out patents occurred very seldom and it was generally frowned upon. This enabled academic scientists to absolve themselves from responsibility for the effects of their findings that the findings might have on other groups in society.

On the other hand, the scientists and technicians who worked on the application of science were mainly employed by industrial companies, whose chief interest was financial profit. Ethical questions about the consequences about the applied research were seldom raised by the employers and the employees were discouraged from concerning themselves with these issues.

Now all this has changed. The tremendous advances in pure science, particularly in physics in the first half of the last century and in biology in the second half, have completely changed the relation between science and society. Science became a dominant element in our lives. It has brought great improvements in the quality of life, but also, great perils: pollution of the environment, squandering of vital resources, but, above all, a threat to the very existence of the human species on this planet through the development of weapons of mass destruction.

The hugely increased role of science in the life of the community brought about by the great discoveries in science has, in turn, resulted in an immense increase in the magnitude of the scientific endeavor; a process of positive feedback in which success breeds further success. Thus, we saw an exponential growth in the number of scientists and technicians.

Sir Joseph, you are now best known for your 1995 Nobel Peace Prize, but you have had a distinguished career as a scientist in more than one field as well. What drew you to science in the first place?

From my early youth on I was fascinated by science. I was reading authors like Jules Verne and other science fiction. From that evolved my passion for science. I wanted science to be applied for the benefit of humanity because of my experiences during the First World War. I saw so much suffering, there were all those terrible diseases, hunger and cold and I myself suffered much. I had a dream as a young child that perhaps one day science would prevent this sort of evil happening in the world. That dream kept me going during those difficult years and later on it became possible for me to become a scientist. This is why I felt science should work for the benefit of humanity.

How the Pugwash Conferences Came About

I resigned from the atom bomb project during World War II as soon as it became clear that the Germans would not make one. I held hope that the bomb wouldn't be used against civilian populations. Japan was already militarily defeated, and yet the atom bomb was used to destroy two cities. When I heard the news about the Hiroshima bomb, I knew already at that time that the thousand times more powerful hydrogen bomb was going to be developed. I was afraid that it would go on like this; there would be a nuclear arms race that would pose a threat to the whole of humanity.

The nuclear era is far from over. US warhead production stopped in 1990, but new design work continues. U.S. explosive testing stopped in 1992 (and in 1990 in Russia) but new "subcritical" tests and computer simulations may yield much of the same information. And sizable arsenals will be retained. Of 85 tons of U.S. weapons-grade plutonium, 47 tons are to be kept for weapons purposes. Russia is the only country still producing new warheads, but the age of its arsenals and its lack of financial resources may nevertheless translate into a sharp decline of its nuclear stockpile in coming years.

—Michael Renner, *Vital Signs 1999*, p. 116

I felt scientists had to do something about it. They know the dangers better than anybody else. The important thing was to be able to talk to our colleagues from the other side of the Iron Curtain, and not just to talk among the British and the Americans. This was impossible during Stalin's era, when no Russian scientist was allowed to come over to talk about such matters. We had to wait until about the mid-1950s after Stalin's death, when the Khrushchev era came. The first time that we met together was actually in a village called Pugwash in Nova Scotia from which we took the name of our movement. Ever since we've been going strong.

A Hippocratic Oath for Scientists

Science affects every walk of life. It may determine the destiny of the whole of humanity. The work of scientists may lead to very great dangers and this is why I feel that scientists should be responsible for what they are doing. If they start a project, they should look ahead, or ask perhaps more senior people, to look into it to see whether it should be done or not.

I suggest that ethical committees, composed of eminent scientists from different specialities, should be set up for the task of examining potentially harmful long-term effects of proposed research projects. The ethical committees should work under the aegis of the national academy of sciences in the given country, but it is essential that the criteria used in the assessment of projects are agreed internationally by academies of sciences, so that the same standards are applied everywhere.

—From Rotblat's Nobel Peace Prize address, 1995

Should scientists themselves impose certain restrictions? I'm particularly concerned with young scientists who start their career. I think they should give a pledge showing that they feel responsible, that they think about what they do. The pledge says: "I won't take any action unless I first examine what might be the consequence of my work. I will not do anything that may cause harm to human beings or the environment."

The time has come for some sort of Hippocratic Oath to be formulated for scientists. A solemn oath, or a pledge, taken when receiving the degree in science, would, at the least, have an important symbolic value, but it might also generate awareness and stimulate thoughts about the wider issues among young scientists.

—From Rotblat's Nobel Peace Prize address, 1995

Human Rights and Responsibilities

I see more and more support for this position. My worry, however, is that there are many scientists who still feel there is no need for them to be concerned. They believe that science is pure. You give to the world your knowledge and it's up to the world to decide how to use it. I call this an immoral attitude. Each of us, not just the scientists, has to be responsible for our deeds.

We talk about human rights, but we speak much less about human responsibilities. If you have rights, you also have responsibilities. I say that the scientist has a particular responsibility because he or she knows well ahead of other people what could be the outcome of a given research project.

Nuclear Deterrence

In my opinion there was no need for nuclear weapons during the Cold War. Some people felt that there was this ideological divide in the world between East and West and that they had to keep the balance of nuclear weapons. Now that we're at the end of the Cold War, I feel even more that there's no need for them.

The United States has formally adopted the policy to eliminate all nuclear weapons. But their actual policy is to postpone this indefinitely. If the most powerful country in the world, the United States, needs weapons for its security, how can you deny such security to countries that are really insecure? The result would be proliferation, as we have seen in India and Pakistan and in 1998 where the danger is very great.

> From the beginning of the atomic age to 1996, the United States has spent a conservatively estimated $5.6 trillion (in 1997 dollars) on nuclear weapons and weapons related programs. Producing the warheads—a total of 70,299 from 1945 to 1990—claimed $417 billion of this sum, while producing and deploying the weapon systems that carry the warhead cost about $3.3 trillion. The US manufactured 6,135 intercontinental missiles, more than 18,000 shorter-range ballistic missiles, and more than 50,000 nuclear capable non-ballistic missiles, as well as 4,680 nuclear-capable bombers and 191 nuclear-powered submarines. Targeting and controlling the weapons absorbed another $845 billion, and a variety of programs to defend against Soviet nuclear weapons took $954 billion. Some $33 billion went to dismantling warheads. Waste management and environmental cleanup efforts have received $67 billion so far, but will require at least $300 billion more during the next 75 years, according to official estimates.
>
> —Michael Renner, *Vital Signs 1999*, p. 116

They say we need the weapons as a deterrent. To me that is immoral because it implies that they would use them. If they really mean not to use them, then it doesn't work as a deterrent.

The only way out is to stop working on nuclear weapons. I agree with Hans Bethe that scientists everywhere, in all countries, should desist from further work on them. If scientists could get together and say: we are all no longer going to work on this because this is something that is causing harm to society, then, of course, there would be no more new weapons and we should be able to live in a peaceful world.

A World without War

Let me give you an example that I have seen in my lifetime. I lived through two World Wars. In both of these wars France and Germany were mortal enemies killing each other by the millions. Now the idea of France and Germany going to war is inconceivable. This evolution gives me hope. If this complete change could occur in about twenty years then why not in other parts of the world? We are moving in the right direction and eventually I think we will be on the road towards a peaceful world.

Hope for the Future

We can no longer afford to have a war because a war could mean the end of the whole of human life on this planet and all the other species. This by itself means we have to do something about it. The young generation is entering a dangerous age, the age where the existence of the human species is no longer guaranteed. This in itself should be a trigger for them to do something to prevent a catastrophe.

Science offers so many ways of preventing the need for war. There is enough food in the world for everybody if properly distributed. Survival used to be the main cause of war because they had to compete for food. This is no longer necessary now. We live in a new world and war must not happen in any case. If we can educate the people in this and foster a feeling of belonging to humanity then I think the future is bright.

My Message to the Youth

What we are doing is not for us. It is for you and your generation. We want to make sure that you will be able to grow up without fear. You can do it if you feel that you are a member of the human species, if you feel that security is no longer only in your nation but is security in the whole world. You must become a world citizen without giving up your loyalty to your nation. You have to extend this loyalty to the whole world. If you do this then the prospects for you to be able to live a life secure, plentiful, and equitable, are very great, but you have to make an effort. It won't come by itself.

References

Rotblat, Joseph. "Science and Humanity in the Twenty-first Century." Nobel Peace Prize address: www.nobel.se/peace/laureates/1995/index.html

YECHIEL BECKER

Science for Peace

Yechiel Becker

Yechiel Becker has been Professor of Molecular Virology since 1971 at the Hebrew University of Jerusalem holding the Carolyn Jane Bendheim Chair of Molecular Virology. Professor Becker has published, since 1955, 390 papers in the fields of molecular virology, antivirals, and dendritic cell research, and in the development of antiviral peptide and DNA vaccines. Since 1987, he has been Editor-in-Chief of the journal *Virus Genes*.

He is a member of Editorial Boards of the journals *IN VIVO, Anticancer Research* (Greece), *Journal of Molecular Medicine* (Germany) and *Acta Virologica* (Slovak Republic). He was Editor of the series *Developments in Molecular Virology, Development in Medical Virology,* and *Developments in Veterinary Virology* (1981–1998), a total of thirty volumes, published by Kluwer Academic Publishers, USA.

Currently he serves as Director of the UNESCO-Hebrew University of Jerusalem International School for Molecular Biology, Microbiology, and Science for Peace, which was inaugurated December 1995. He has organized the *International Symposia on Science for Peace in Jerusalem* (1995, 1997, and 1998) and *International Workshops on Different Aspects of Molecular Biology and Microbiology* (1995, 1997, and 1999). The *Jerusalem Statement on Science for Peace* (1997), the *Eilat Statement on Science for Peace* (1997), and the *Jerusalem Statement on Science for Peace and the Need to Eliminate Biological Weapons* (February 1998) were issues.

Professor Becker, you are a molecular microbiologist who got involved in the global peace process. This seems, at first blush, a most unlikely combination of passions. How did you come to that path and what does it say about the responsibilities of science in the social milieu?

As you know, the Oslo Agreement was signed in 1993 by Shimon Peres, Yitzhak Rabin, and Yasser Arafat. In January 1994, while I was visiting the headquarters of UNESCO-Paris, I was asked by Prof. Claude Rosenfeld whether I had any ideas for UNESCO that may support the peace process. My suggestion was to set up an International School for Molecular Biology, Microbiology and Science for Peace (ISMBM) in collaboration with UNESCO. That was based on my experience during the Gulf War when we in Israel were left alone and no one scientist stood up to say that chemical, biological, and any other weapons of mass destruction are not to be used against civilized people.

Prof. Adnan Badran, UNESCO's Assistant Director General for Science, a Jordanian, supported my proposal on the spur of the moment. He gave the green light and one year later signed the agreement for the International School with Prof. Y. Ben Arieh, Rector of the Hebrew University of Jerusalem, and myself, as the Director of the International School.

The ideas of Science for Peace were to support the concepts of UNESCO's Culture of Peace Program. I felt that Science for Peace could be more easily accepted because the language of science is universal. The inauguration of the

UNESCO-HUJ ISMBM and the first International Symposium on Science for Peace took place in December 1995.

The International School had accepted a mission to develop scientific contacts in the Middle East and with universities and organizations around the world. Of course the International School is without walls, a place where people can meet and exchange views, with activities in connection with a network of schools in many countries.

An international Science for Peace Network is under development to organize Science for Peace and scientific workshops for scientists in European and Middle East countries.

My Inspiration

It's a tragedy that Prime Minister Rabin was killed by one of our own. He was the leader during the Six Day War and to us he was a hero. After that war, people said that Rabin was waiting for a telephone call from President Gamal Nasser of Egypt to settle the differences and to discuss peace. That was the great hope of the Israeli people. At that time no one thought about territories, only about peace. The telephone call did not happen. The concept of peace was probably on Rabin's mind and he inspired me by saying that strategically the time for peace had come.

In the memorial to Rabin at UNESCO-Paris it is inscribed: "Since wars begin in the minds of men it is in the minds of men that the defenses of peace must be created." We are trying to help the building of the defenses of peace.

> The PEACE program is a cooperation of Palestinian and European Universities, launched in 1991, predating the Oslo agreement by two years. It is supported by UNESCO and the European Commission. Higher education in the West Bank and Gaza includes eight universities. There are sixteen community colleges offering two-year diplomas in technical and commercial fields. The universities offer programs in arts, sciences, commerce, economics, engineering, agriculture, law, pharmacy, medical professions, nursing, education, tourism and hotel management. The total enrollment for 1997/98 was 52,427 students, mostly at the undergraduate level. There were 1,678 graduate students, mainly at the two major universities of Birzeit and in Nablus. The total number of academic staff is 2,215.
> —UNESCO documentation on the PEACE program, 1997

The Jerusalem Statement on Science for Peace (1997)

The *Jerusalem Statement on Science for Peace* was presented to the participants of the *Second International Symposium on Science for Peace* that was held in

Jerusalem in January 1997. We felt that it is crucial to have an open communication between scientists and no political pressures or demands from governments on scientists. We thought that Science for Peace is the right ethical approach for the future. Scientists must work for the benefit of humanity, not just for the benefit of themselves. When they do research in biology or biotechnology, the fundamental sciences of life, they should think of how to improve life for people in their region and elsewhere.

The concepts of Science for Peace of the Jerusalem Statements were incorporated into the UNESCO *Declaration on Science* and the *Science Agenda-Framework for Action* (UNESCO, Paris, 2000).

A Science for Peace Oath

Participants of the *Second International Symposium on Science for Peace* (1997) also suggested a Science for Peace Oath for young scientists to take when they graduate. The Science for Peace Oath will remind the scientists that their research must be used for the benefit of humanity and the environment. There is an urgent need to convince educators and universities to include the Science for Peace concepts when teaching young scientists, and the Oath for Scientists should be included in the graduation ceremony of scientists, similar to the inclusion of the Hippocratic Oath in the graduation ceremony of medical school students.

The University for the Middle East

An additional idea that was developed by the participants of the 1997 *Symposium of Science for Peace* was the University for the Middle East. We wanted to find ways of bringing together scientists from the Middle East countries who are uncooperative even though formal peace was signed between the countries. There are barriers that do not allow scientific collaboration.

The *Jerusalem Statement on Science for Peace* (1997) suggested "the establishment of a world class international institution of higher learning and research, open to all students, educators, researchers and administrative members, without regard to nationality, country of origin, political beliefs, religious faith or gender." The concept of the University of the Middle East was supported by UNESCO and I hope that in the near future it will materialize.

*The Second Jerusalem Statement on the Need to Eliminate Weapons of Mass
Destruction and to Prevent the Misuse of Biological and Chemical Sciences
(1998): National and International Support*

I issued this statement on 2 February 1998 when a new phase of the Gulf War
against Iraq was imminent. The statement was straightforward and simple:
"Never more biological or chemical war."

This statement was distributed by the American Society of Microbiology and
the American Society of Biology. The American Society of Microbiology has
45,000 members and the American Society of Biology about 3,000. Some 270
scientists from 27 nations supported me, among them leading microbiologists.

Within Israel the response was also from citizens who wrote to me that what
we do is important. I received a supportive letter from a group of citizens in the
North of Israel. We cannot expect the whole population to respond or understand that they have the right to peace, but the statement was broadcast on the
local radio. Once is not enough and we have to repeat it and I hope that we will
continue. Unfortunately there is a country in the Middle East which, despite
the Gulf War, still has biological and chemical weapons, as has been stated by
Richard Butler, former Director of UNSCOM.

Eventually, I intend to collect all the information derived from our Science
for Peace activities in book form and to try to publish it as part of the effort to
prevent biological or chemical wars. For many years Sir Joseph Rotblat and the
Pugwash Conferences were working mainly to eliminate atomic weapons. Now
the biologists should pool their strength together to support a movement
against biological and chemical wars.

Bioterrorism

What will happen with engineered bacteria or viruses that are used in biological weapons? In 1999 Professor Donald A. Henderson had started a Center for
Civilian Biodefense Studies at John Hopkins University. He was responsible, in
the World Health Organization (WHO), for vaccinating the world population
against polio and smallpox, and he is worried about the possibility of smallpox
virus falling into the hands of terrorists. Bioterrorism is a major threat. We
have to develop new anti-viral drugs and vaccines because the available ones
may be inefficient against engineered microbial agents.

> *Bacillus anthracis,* the organism that causes anthrax, derives its name from the
> Greek word for coal, *anthracis,* because of its ability to cause black, coal-like cuta-
> neous scars. Anthrax infection is a disease acquired following contact with in-
> fected animals or contaminated animal products or following the intentional re-
> lease of anthrax spores as a biological weapon.

In the second half of the last century, anthrax was developed as part of a larger biological weapons program by several countries, including the Soviet Union and the U.S. The number of nations believed to have biological weapons programs has steadily risen from 10 in 1989 to 17 in 1995, but how many are working with anthrax is uncertain.

Perhaps more insidious is the specter of autonomous groups with ill intentions using anthrax in acts of terrorism.

—Center for Civilian Biodefense Studies

At the beginning of the nineteenth century, the average life expectancy was about forty-five years. Today, it is almost doubled, and this is because scientific and medical research led to the control of infectious diseases, microbial epidemics, and various other medical conditions. The need to eliminate weapons of mass destruction, especially biological weapons, is as urgent as the need for food, for what will happen if food will be contaminated by the misuse of microorganisms? The peoples of the world have the right to peace and they must demand that weapons of mass destruction, especially biological weapons, will be eliminated.

Vaccines for Peace

The concept of Vaccines for Peace was developed by the German scientist Prof. Erhard Geissler in Berlin and by colleagues who offered this concept to the WHO and to the committee for the Biological Weapons Convention. The idea is that once you have the vaccines against biological weapons, the population will be protected and the weapons will be useless.

There are currently no atmospheric warning systems to detect an aerosol cloud of anthrax spores. The first sign of a bioterrorist attack would most likely be patients presenting with symptoms of inhalation anthrax. An analysis by the Office of Technology Assessment of the U.S. Congress, estimated that 130,000 to 3 million deaths could occur following the release of 100 kilograms of aerosolized anthrax over Washington D.C., making such an attack as lethal as a hydrogen bomb.

The U.S. licensed in 1970 a human anthrax vaccine mandated for use in all U.S. military personnel. In studies with monkeys, inoculation with this vaccine at 0 and 2 weeks was completely protective against infection from an aerosol challenge at 8 and 38 weeks, and 88% effective at 100 weeks. However, U.S. vaccine supplies are limited and U.S. production capacity is modest. There is no vaccine available for civilian use.

—Center for Civilian Biodefense Studies

Why should we be under the threat of biological weapons? We should stand up and say: "We will not tolerate that!" The media is more interested in creating a big story about what could happen and how important it would be to have antibiotics and to prepare ourselves without really explaining how biological weapons will affect children and adults.

Atomic weapons are banned by the International Convention because we know the outcome of using them. Chemical weapons are banned by the Chemical Weapons Convention since World War I. Biological weapons are banned since 1925 but the Biological Weapons Convention (BWC) is not yet finalized. However, there are nations that signed the Convention Agreement and yet, under its cover, developed and stored dangerous biological weapons. The agents are very dangerous because they have the capability of multiplying and mutating in infected people and animals. One cannot control mutations in living microorganisms, and mutations may lead to the development of more dangerous pathogenic microorganisms that may cause a worldwide epidemic.

A Safe Alternative for Producing Vaccines for Immunization of Populations

Today's vaccines are produced in factories where the pathogenic viruses or bacteria that are the infectious agents are grown on a large scale and then inactivated to become vaccines. This is still the same idea as when Jonas Salk developed the polio vaccine or Pasteur's idea when he produced the vaccine against rabies. Once you have a facility to grow infectious agents in large quantities for vaccines, the same facility can be used for biological warfare.

> Two decades ago, only 5 percent of children in developing countries were immunized against six child killers: measles, polio, diphtheria, whooping cough, tuberculosis, and tetanus. Today, thanks to a global effort led by UNICEF and its partners, close to 80 percent of the world's children are immunized. This incredible achievement now saves the lives of 3 million children every year. However, 3 million children die every year from diseases that are entirely preventable. Still, 60 percent of the children are not immunized in almost 50 nations. While one child can be fully immunized for $17, every $1 spent on immunization saves society up to $29.
> —UNICEF USA

How could one eliminate the need for such factories? There is an ongoing discussion with countries that have the BW facilities to transfer their BW laboratories to the production of useful products for human use. How can they be convinced to convert the facilities to producing materials that are beneficial to the people?

My suggestion is actually to use new vaccine technology that has been developed requiring only synthetic peptides derived from bacteria or viral proteins rather than from the infectious organisms. These peptides are synthesized in open laboratories and we no longer need the infectious agents since we have the nucleotide sequences of the nucleotide sequences of the genomes of the pathogenic microorganisms. By computer analysis it is possible to define in the microbial proteins the antigenic domains, and the synthetic antigenic peptides can be chemically synthesized and used as vaccines to protect humans or animals against pathogenic microorganisms.

If this approach works then the countries with factories that grow infectious agents are doomed to be condemned by the international society. There will be no need to have inspections of conventional vaccine production as required by the BWC. Such inspections are the major reason why the agreement on the Biological Weapons Convention was not finalized. Industries do not want to have BWC inspectors because they are protective of their patents. With synthetic peptide vaccines it will be possible to produce antimicrobial vaccines at a relatively low cost.

My concept is that the synthetic peptide (or DNA) vaccines be introduced into the human skin as peptide/DNA vaccines to eliminate the use of syringes since immunization by the skin route induces the cellular immune response against a putative pathogen within two days. When the world population has been immunized against man-made and natural infectious agents in all parts of the world, there will be no use for biological weapons. I feel that the public should stand up and say to its leaders: "We will not tolerate threats of bioweapons. We have the right to live in peace."

It is a great disappointment to know that still some countries continue to develop biological weapons, which, if not eliminated, can cause a global catastrophe. BW are not geographically limited like chemical or atomic weapons. Pathogenic anthrax bacteria do not have borders and a BW catastrophe of a large scale must be prevented. The responsibility of the UN and the scientific world is to lead the demand to finalize the BW convention and to demand its implementation from the world leaders.

Regional Cooperation: The Next Step

We had organized in 1997 in Eilat an *International Workshop on Virus Diseases in the Middle East and Neighboring Countries.* At the end of the workshop the participants had issued the *Eilat Statement on Science for Peace* and the need to develop a Middle East Scientific Network, containing the idea to develop a Middle East Scientific Research Network. We had participants from Jordan, Egypt, Palestinian Authority, Europe, and the U.S. Such a Scientific Network will enhance the development of collaborative scientific projects for the benefit of all Middle East and Mediterranean countries.

This concept is based on my experience in the European Common Market. In 1971 I was invited by Raymond Graps to serve as scientific advisor on agriculturally important diseases caused by animal viruses. The European Commission Directorate General VI was responsible for the organization of collaborative research on virus diseases for the scientists in the nine member states and in the two associated states, Portugal and Spain. I undertook the challenge of developing ways to help the cooperation between scores of laboratories in many projects for nearly fourteen years.

The scientific collaboration was strengthened by developing scientific cooperation on virus diseases of chickens, pigs, and cattle. The results of these research programs had led to the development of new diagnostic measures that allowed the EEC administration to prevent the import of diseased animals to the EEC member states. The same approach was used to eradicate diseased animals from the EEC member states and strengthened the farmers in the EEC nations.

Granada II: A UNESCO Project for Collaboration between Palestinian and Israeli Scientists

Prof.Federico Mayor, at the time Director General of UNESCO, had decided in 1998 to develop the Granada II project by contributing $1 million for ten projects in education, science, and culture for collaboration between Palestinians and Israelis. The Palestinian-Israeli Steering Committee approved three collaborative scientific projects on human genetic disorders, virus diseases of humans, and the prevention of rabies virus infection in animals.

The scientific cooperation in the Granada II project made it possible for the UNESCO-HUJ ISMBM and Science for Peace to organize in 1999 the *UNESCO Workshop on Science for Peace and Regional Scientific Cooperation in Molecular Biology, Microbiology, and Biotechnology* in Jerusalem. The workshop was attended by Palestinian scientists and students from Al-Quds University and Bethlehem University, and collaborating Israeli scientists.

Building Bridges for Peace: The Story of Palestinian Students Studying Microbiology at the Hebrew University of Jerusalem

One of our first ideas of the UNESCO-HUJ ISMBM at the Hebrew University of Jerusalem was to open the door, literally, for a Palestinian student, from Gaza, to enter the Hebrew University to study for a master's degree in microbiology. For this I received financial support from Mayor at the end of 1996.

Around that time there was a very difficult period of terrorist attacks against Israelis. This student kept calling me from Gaza apologizing for these deeds. He felt that we have to live together and hoped that when he will graduate he would be in a position to be listened to, so that Palestinian science and peace could be fostered.

When he came to us he was working at Bir Zeit University (Palestinian Authority) where he had graduated with a B.Sc. degree. He had the credentials that enabled his acceptance to the Hebrew University of Jerusalem since the International School is a door opener but the Hebrew University authorities decided on the applicant's credentials.

The Institute of Microbiology's teaching committee accepted him with great enthusiasm because he was the only Palestinian student from Gaza applying to study at the Hebrew University of Jerusalem. However, first he had to receive the approval of the Palestinian Authorities. He went to the Dean of his university and to the President who told him to go to the Minister of Higher Education, who gave her approval. From the Israeli Foreign Office he received an Israeli visa in a Palestinian passport, perhaps the first, and he came to our university.

He was immediately accepted. Dormitories are always hard to come by, but he received a room. I think that he is a very special person, a poet, a storyteller and an excellent young scientist. In November 1998, he gave a talk at the Symposium on Science for Peace. Usually there are about sixty participants, half from Israel and half from other countries. He told his story, titled "From Al Azhar University in Gaza to the Hebrew University in Jerusalem." He was amazed to have been accepted by the Israelis. Some of his best friends are Israeli students who have been in the military. He could not reconcile his experiences with the Israeli military on the one hand, and the kindness of the people at the university on the other.

This paves the way, and delivers a message to other Palestinian students that it is possible to study together at the Hebrew University. Currently, two more Palestinian students were accepted. One is a Ph.D. graduate in virology from a university in the United States who has been accepted at the ISMBM as a postdoctoral student, and the other is studying microbiology for a M.Sc. degree at the Institute of Microbiology and works part time as a technician in a Palestinian hospital in Jerusalem.

Two additional Palestinian Ph.D. students received short-term fellowships from the international school. The three Palestinian students were financially supported by grants to the UNESCO-HUJ ISMBM provided by UNESCO-Paris and later by the Fritz Thyssen Stiftung in Koln, Germany.

Toward Comprehensive Science for Peace in the Middle East

This is how I see that peace can emerge. I sympathize with the Palestinian students and scientists because when I started my studies in 1951, the Hebrew University of Jerusalem could not use its Mount Scopus Campus that was in Israeli territory, an enclave in the Jordanian occupied territory. We couldn't use the facility nor the library and everything had to be built from the start. During the period 1949–1965 our medical faculty was in a temporary building, close to the walls of the old city in Jerusalem. We have since built an excellent university with teaching and research facilities. I'm sure that in the near future the Palestinians will have a strong cadre of professors, teachers, scientists, and students. I am pleased to have the opportunity to give a hand. We are doing this for many other countries, so why not for our neighbors?

Many Israeli and Palestinian scientists from Israeli and Palestinian universities share the view that scientific collaboration is a must. This view is presented in the Jerusalem Statement on Science for Peace. We still have a barrier to overcome, to develop free scientific cooperation with scientists from Egypt and Jordan. UNESCO should continue to provide help because with their support the Palestinian and Israelis scientists are able to collaborate. We have scientific centers in Israel that are open to all scientists. All the universities are developing scientific collaborations with our neighbors, a process that needs long-term support. I look forward with hope that in the near future the UNESCO International School of Science for Peace will be developed in the countries of the Middle East.

References

Becker, Yechiel, and V. Kouzminov. 1997. *Science for Peace.* International symposia, Hebrew University of Jerusalem. UNESCO, Venice Office (ROSTE).

PEACE: Palestinian European Academic Cooperation in Education, www.unesco.org/general/eng/programmes/peace-palestine/

IV

Science for the Twenty-first Century

ISMAIL SERAGELDIN

The Cry of the Eight Hundred Million

Until 2000, Dr. Ismail Serageldin was Vice President of Special Programs at the World Bank. He chaired the Consultative Group on International Agricultural Research (CGIAR), the Global Water Partnership (GWP), the World Commission on Water for the Twenty-first Century, and the Consultative Group to Assist the Poorest (CGAP).

Dr. Serageldin led the bank's program in the promotion of cultural heritage in developing countries. He also designed and managed a wide range of poverty-focused projects in developing countries, and from 1993 to 1998 he was Vice President for Environmentally and Socially Sustainable Development. In 2001 he was appointed the first director of the new Library of Alexandria in Egypt.

He is a prolific writer on such topics as economic development, human resource issues, and the environment. Among his recent publications are *Nurturing Development: Aid and Cooperation in Today's Changing World* (1995), *Sustainability and the Wealth of Nations—First Steps in an Ongoing Journey* (1996), "Biotechnology and Food Security in the 21st Century" (*Science,* July 1999), and *Social Capital—A Multifaceted Perspective* (with Partha Dasgupta, 2000).

Dr. Serageldin, you hold the unusual position of bridging cultural imperatives and technical ones, from within the massive edifice of the World Bank. What are some of your foremost passions and commitments?

As Vice President for Special Programs at the World Bank, I deal with a number of issues, for example, culture and development and the notion of cultural identity. With globalization and the increasing feeling of alienation, this becomes an important part of how to build a sense of self-confidence, a sense of continuity with the past and at the same time, openness to the future. That's a central part of dealing with the development paradox. That's something I do for the World Bank.

The Consultative Group on International Agricultural Research (CGIAR), is the most amazing and astounding success story of international collaboration in science to promote the well-being of the poor and for future generations. It is a very large institution; it's nearly thirty years of age, it involves over 1,100 international agricultural scientists, 9,000 employees with a $340 million annual research budget. There are 16 international centers dealing with promotion of sustainable agriculture for food security, poverty reduction, and environmental protection in the developing world.

> The external debt of developing countries rose to $2.2 trillion in 1997. Developing countries spent $269 billion servicing their debt, compared with $191 billion in 1990. The outright cancellation of debt for countries in Africa could save the lives of 7 million children annually by 2000 and provide 90 million young women with access to basic education. For the United States this debt forgiveness amounts to roughly the cost of building two B-2 bombers.
>
> —Lester R. Brown, *Vital Signs 1999,* p. 16

To complement that, I chair the Consultative Group to Assist the Poorest. That's an effort to see how we can assist in reaching more poor people and reach deeper into the poverty profile. How can we empower the poorest, especially women? There are initiatives like the Grameen Bank in Bangladesh, Self Employed Women's Association (SEWA) in India, and FINCA International. Micro finance brings human dignity and enables these poor women to take charge of their own destinies.

Finally, I believe that one of the major issues that we haven't dealt with properly in the past has been fresh water. That is why I'm also involved with the Global Water Partnership and chair the World Commission on Water for the Twenty-first Century. The latter activity prepared a report for World Water Day 2000 that got us to change our attitude in the management of water. Unless we change the way we are approaching the issue, the wars of the twenty-first century will be fought over water much as the wars of the twentieth Century have been fought over oil. All of these are long-term issues; they are profound; they deal with the human condition; they harness the best of science and knowledge for the benefit of the poor and future generations.

The Cry of the Eight Hundred Million

There are about 1.3 billion people who live on less than a dollar a day. There are 3 billion people on the planet who live on less than $2 a day. Two billion people have no access to electricity; about 1.4 billion have no access to clean water; almost 3 billion have inadequate sanitation. We have close to eight hundred million people who are chronically malnourished, outright hungry, which is of course a disgrace and an insult to our common humanity. We are just celebrating the fiftieth anniversary of the Universal Declaration of Human Rights. How about the denial of the most basic of all human rights, for somebody to eat and not to see their children stunted in their growth because of inadequate nutrition? These are the challenges of today!

> Roughly one out of five people in the developing world, primarily women and children, goes hungry every day, lacking enough calories and protein to satisfy basic body requirements. The greatest concentration of these less fortunate people is in South Asia and sub-Saharan Africa. More than 20 percent of South Asia's population is starving, with rates as a high as 30 percent in Bangladesh and 70 percent in war-torn Afghanistan. Some 200 million—over one out of five people—go hungry in India alone. South Asia is home to about half of the world's starving children.
> —Brian Halweil, *Vital Signs 1999*, p. 146

What Can We Do for the Poor?

We know what works and what doesn't work. There's no single magic bullet, but you have to start with sound macro-economic policies. Major investments in human resources: education, nutrition, and science. Investment in social capital enabling people to have the social cohesion that is necessary for societies to function. There is no society that has successfully developed without focusing on agriculture and getting its agriculture sound because otherwise there would be no reduction in poverty, no protection of the environment and no food security in the world. Science has a very big role to play in that area, which is what the CGIAR is involved with.

We have to focus on removing discrimination. Women are systematically discriminated against and we also know that the education and empowerment of women is at the root of almost every positive social development including fertility reduction, reduction in infant mortality, improvement in child nutrition, and improvement in almost any social indicator.

These are all the tools that we have. Empowerment of the weak, involvement in human capital and in social capital, mobilization of science for agricultural transformation in order to promote environmental protection, poverty reduction, food security, and sound macro economic policies; that's the package.

The Role of Women

I think it's basically the patriarchal power order that exists in most societies that has discriminated against women systematically. There has been a double standard in society and these cultural norms run deep and die slowly. No society can advance with half its citizens deprived from flourishing to their full potential. What we're discovering in the case of women is that in fact they are the vectors of social transformation of society. They are the ones who create social cohesion. They heal the wounds in postconflict situations. They reach out and create solidarity groups of the grass roots; they educate and articulate the values for the next generation. Thus the education and empowerment of women is central to almost everything we do.

The Missing One Hundred Million

Nobel Laureate Amartya Sen wrote a piece a few years ago that shocked the world. It's called "100 million women are missing." He took age specific, sex specific mortality rates, and showed that in the Indian subcontinent, if world average figures had prevailed, there should be 100 million women more than there actually are. There was systematic discrimination against girls and not necessarily in terms of formal infanticide, but in terms of inadequate nutrition, inadequate attention to health care, and with females less likely to be educated.

All of this is happening in other places as well. But the heart of the matter is an attitude. If we genuinely believe in human rights, if we genuinely believe that women's rights are human rights, they're indivisible. It's the responsibility of everybody to denounce this discrimination at every possible opportunity.

Food Security

Food security is not self-sufficiency of food production by every nation. Food security means that at all times, all people will have access to adequate amounts and quality of food that is within reach, economically, socially, and environmentally. Also physiologically, that they are able to absorb it. As you know there are children who are sick with malaria or diarrhea and they are unable to absorb food even when it's available. Food security is all of these dimensions together.

> Undernourished children are at a higher risk from most infectious diseases—including pediatric killers like diarrhea and pneumonia. About half the deaths in children worldwide are associated with malnutrition—more than 6 million deaths each year. Since inadequate food intake in childhood leads to permanent mental and physical stunting, underweight children today represent a potent barrier to a nation's social and economic progress.
> —Brian Halweil, *Vital Signs 1999*, p. 146

How do you achieve that in all societies? It is a complex of issues; it is not just production, but it's also access. If it were just production, there would be no hungry people in the United States, the world's largest exporter. There are estimates of over twenty million hungry people in the United States, most of them children.

It's also a matter of how it is produced: whether it is produced sustainably or are we eroding the soils and chopping the forests? Today, 10 percent of the world's agricultural production is unsustainable because it's over drawing underground water reserves faster than the recharge rates.

Urbanization and Rural Small Holder Farms

Urbanization is growing incredibly rapidly in the developing world. In India alone, the increment in urban population in the next thirty years is going to be more than twice the population of France, Germany, and the United Kingdom combined. The best thing you can do for the urban poor is to reduce the price of food. Therefore, more effective rural development that empowers smallholder farmers, which produces surpluses and improves the conditions of the rural poor, and provides adequate cheaper food would translate directly into increased income for the urban poor. The solution again comes from the transformation of the rural world. Seventy percent of the world's poor are still in rural areas.

As the economy grows, pressures on the Earth's natural systems and resources intensify. From 1950 to 1997, the use of lumber tripled, that of paper increased sixfold, the fish catch increased nearly fivefold, grain consumption nearly tripled, fossil fuel burning nearly quadrupled, and air and water pollutants multiplied several fold. The unfortunate reality is that the economy continues to expand, but the ecosystem on which it depends does not, creating and increasingly stressed relationship.

—Lester R. Brown, *State of the World 1998*, pp. 3–4

The Role of Science

The way to reduce the price of food is to increase production, by intensifying production at the small holder level. That's where science comes in. There is a shift in the research paradigm going on, which is to contextualize traditional crop-focused agricultural research. You look at the micro ecology, which means that we try to see how it fits best in the particular conditions. It means recognizing that the small farmer who holds anywhere from half a hectare to two hectares of land never grows a monoculture. He or she has some food crops, some cash crops, and maybe a couple of tree crops; some livestock, some chickens. We need to find ways in which to capture the synergies of that particular farming system.

Capturing the Synergies

One of the CGIAR centers, the Nairobi-based International Centre for Research in Agroforestry (ICRAF) has done a lot of very interesting work in agroforestry. They screened some two thousand different types of trees and selected four trees: *Gliricidia, Leucaena, Sesbania,* and another one. Those are multipurpose, leguminous trees that are nitrogen fixing, so they help other crops next to them. Second, they grow incredibly fast and the women can harvest them for firewood. The animals love the leaves as fodder and it's very nutritious for them. That's an example of a multi purpose tree that was designed not for timber production and cash production, but in terms of thinking of the complex farming system of a small holder farmer.

Others produce "green manure" that can be used as fertilizer by being chopped up and plowed under. Then you find techniques that involve, for example, proper sequencing of crops. Aquaculture ponds not only produce fish for food and sale, which is an additional source of income, but also help better manage soil moisture on their small plots. Corralling livestock on fields before planting enables them to collect organic matter all around and then depositing it again in terms of manure that increases returns.

The Gender Dimension

Part of the contextualization is the socioeconomic characteristics and among those there's the gender dimension again. In very large areas of Africa the men grow the cash crops and the women grow the food crops. The farmers receive the extension services and the credit lines. The women actually produce 80 percent of the food, but earn about 10 percent of the wage income and own 1 percent of the land.

Agriculture in the United States

For the United States, it's not the romantic return to the family farm that is a solution; it's more the movement toward precision farming and the better environmental management of their farming system and the more appropriate utilization of fertilizer. Eliminate the leaking of nitrate into the soil, and ensure better protection of the topsoil, a number of other issues of that kind. Most Americans (not all) are way beyond the family farm given the overall structure of the economy. Developing countries like Bangladesh, India, or countries across Africa are at a starting point where 70 to 80 percent of the population is rural and lives in small farms.

The Green Revolution

Whatever critics of the Green Revolution talk about, without it in the past thirty years, to produce the same amount of food as today, we would have had to bring under cultivation three hundred million hectares of land as a very conservative estimate. That is an area the size of India. That's more than the total arable land of the U.S., Canada, and Brazil combined. Can you imagine how many more forests would have had to be chopped down? How many more habitats would have been destroyed? How many more species we would have lost? How much more biodiversity would have been extinguished? How much more soil erosion would have occurred?

The Green Revolution certainly staved off the potential famine that was predicted in widely published books like *Asian Drama* by Gunnar Myrdal and *The Population Bomb* in 1968 by Paul R. Ehrlich. Myrdal's prediction at the time was that India could not then feed its 450 million people, and it was going to have at least 750 million in the subsequent generation. Today, with almost 1 billion people, India provides caloric coverage for 950 million, largely on the same amount of land and water as in 1975, and that includes a 33 million ton surplus in terms of grain. The malnutrition of 200 million people in India is very much the access side of the question, not the production side. It's the same with the famines in the world that have occurred in former Biafra, Somalia, and the

Sudan. Those are famines that are really associated with civil strife, thus preventing access.

Increasingly, the CGIAR is now focusing on the socioeconomic characteristics of poverty reduction. Hence the focus on the small holder farmer and not just on the production levels or the income levels. All this was an evolution of thinking over a twenty-five, thirty-year period.

The New Abolitionists

I came from a fairly privileged family in Egypt where I was born and raised. However, I always went to public schools and was very aware of the huge gaps in income in the society in which I lived. I always had a sense of social responsibility, that greater equity was required between people. Most individuals, I feel, have an innate sense whether something is fair or unfair. That is why my work has been in the development field, and I have been driven to try to fight extreme poverty and to respect our common humanity and give each person a fair chance in life.

In the last century, people looked at the condition of slavery and said it was unconscionable and must be abolished. They were known as the abolitionists. They were not deterred by the fact that slavery had existed from the beginning of time, that enormous economic interests were arrayed against them. They took it from a position of moral outrage and they said, this is just not acceptable; it has to be abolished.

Today it is unconscionable to have eight hundred million human beings denied the most basic of all human rights that is access to food in adequate quantities. Hunger in today's world is unconscionable and it must be abolished. All of us, with the same moral outrage, from a sense of our common humanity, must denounce that and consider ourselves the new abolitionists. You cannot have inequities of that magnitude and complacently watch the gaps growing wider between the rich and the poor, both within countries and between countries, and sort of shrug your shoulders and say, well, there's nothing I can do about it. Each of us has a responsibility to act on it and each of us must act on it.

References

Ehrlich, Paul B. 1968. *The Population Bomb.* New York: Ballantine.

Myrdal, Gunnar. 1968. *Asian Drama: An Inquiry into the Poverty of Nations.* New York: Pantheon Books.

M. S. SWAMINATHAN

The Soul of Sustainability

Born in India, Dr. M. S. Swaminathan trained as a geneticist at the University of Cambridge, United Kingdom, where he received his Ph.D. in 1952. Dr. Swaminathan returned to India, and became widely recognized as the scientific architect of the Green Revolution in India in the 1960s. As a cytogeneticist and Administrator of the Indian Agricultural Research Institute he developed genetically improved crops such as wheat, rice, and potato during the period 1947 to 1972. As a result, the total Indian wheat yield doubled in four crop seasons from 12 to 23 million tons. He has held many important posts both in India and abroad, is a Fellow of several national science academies, and has received numerous prestigious international awards.

His work in crop genetics and sustainable agricultural development in India and in the Third World earned him the Ramon Magsaysay Award in 1971 and the first World Food Prize in 1987. He is currently Chair of the M. S. Swaminathan Research Foundation, which he described as "an adventure in integrating modern science with traditional wisdom." The Foundation's work started in the field of agriculture, but it has also branched out into projects such the Information Village and the Children for Happiness Movement.

Dr. Swaminathan, you have been one of the key players of the Green Revolution. Explain for us the genesis of the Green Revolution and what agriculture now requires to make it sustainable.

The term *Green Revolution* was coined in 1968 by William Guad of the U.S. Department of Agriculture. At that time the Wheat Revolution had started and the Rice Revolution was beginning. These were large quantum jumps in productivity. For example, in India in about four year's time from 1964 to 1968 we increased wheat production of the same order as in the previous four thousand years. Green is the color of chlorophyll, which absorbs sunlight, the only primary producer of energy. In that sense, agriculture is the largest solar-energy harvesting enterprise in the world.

A number of problems with the Green Revolution were pointed out by environmentalists, social scientists, and economists. For example, in 1972 the price of petroleum products went up steeply, including fertilizers. Economists felt that the Green Revolution that is based upon synthetic mineral fertilizers may not be sustainable because those prices will go up. Social scientists said that here are many resource-poor farmers who will not be able to buy those market-purchased inputs. Traditionally they were only growing farm-purchased inputs; suddenly you had a transition to market-purchased inputs.

There was also the problem of gender. The women were getting displaced, there was mechanization, and the woman's traditional role was eroded. The women were the seed producers while large seed companies are male dominated. With all of these issues there was a feeling that the Green Revolution in the longer term may be more of a social and ecological disaster than a blessing.

The Evergreen Revolution

The Green Revolution meant a quantum jump increase in productivity. For example, in India we are now producing 72 million tons of wheat from about 20 million hectares of land. In 1968, before the onset of the Green Revolution, we were producing 12 million tons of wheat from 14 millions hectares of land; that means less than 1 ton per hectare. If we are now to produce 72 million tons of wheat at the 1965 yield level we would require a minimum of 75 million hectares of land, not 20. Where will the other 45 or 50 come from? The remaining forests would have disappeared. In a way we had a forest-saving agriculture already in the sixties.

The *Evergreen Revolution* is what I call a growth in productivity that is also economically and ecologically sustainable. The key lies in those technologies that are not generic but situation-specific in terms of ecology, environment, social requirements, duration of the crop, the quality of eating, how the food is prepared, and so on. These are all local adaptations.

Agriculture and Biodiversity

> Today nearly 1,000 major agricultural pests—including some 550 insect and mite species, 230 plant diseases, and 220 weeds—are immune to pesticides, a development almost unheard of in the middle of the 20th Century. As modern agriculture leans heavily on pesticides, spreading resistance threatens to increase pest-induced crop losses and weaken food security.
> —Brian Halweil, *Vital Signs 1999*, p. 124

Genetic homogeneity enhances genetic vulnerability to pests and diseases. Even as early as 1968, the very year when the term Green Revolution was coined, I warned against monoculture of the same variety. China and Japan contributed the seminal genes in our wheat and rice. Using those we produced a very large number of varieties that are locally adapted, so we kept a large genetic diversity.

> Pesticide resistance drives the paradox of modern pest control: despite enormous increases in pesticide use since the middle of the 20th Century, the share of crops lost to pests has not changed substantially and has even increased in some regions. In the United States, for example, while pesticide use jumped 10-fold from the 1940s to the 1990s, the share of crop lost to pests actually rose from 30 to 37 percent.
> —Brian Halweil, *Vital Signs 1999*, p. 125

The plants were shorter so that if it gets more nutrients and more water it will not fall down because it becomes top-heavy. In the case of wheat we introduced the Japanese Norin dwarfing genes with the help of Dr. Norman Borlaug of Mexico. The other gene is the Chinese Dee-gee-woo-gen that we used for dwarfing rice.

Participatory Breeding and the Genetic Cord

Women farmers in particular maintain a lot of diversity, in-situ on-farm so to speak. They are conserving large numbers of varieties for the public good. These varieties generally are low yielding so you can say that they are doing a sacrifice by still cultivating them.

The government does not compensate them for this. That is why we have been pressing for community conservation and Gene Banks to be public funded. In-situ conservation is publicly funded through national parks, protected areas, biosphere reserves, and heritage sites. This kind of conservation practiced by millions of farm families, particularly women, is the soul of intraspecific variability. For example, the Gene Bank in the International Rice Research Institute in the Philippines has nearly a hundred thousand rice varieties that come from these small farms.

This is a kind of genetic cord. Through participatory breeding you keep their own varieties; introduce a few genes by working with them, by collaborative breeding. Suppose the local variety is highly susceptible to a particular disease, blast in rice. Then you bring some sources of resistance and also try to improve them by cross-mating among the local varieties. In other words, the very soul of sustainability is highly location-specific agriculture.

The Soul of Sustainability

The *Soul of Sustainability* is adaptation to local conditions. When I say local conditions, I do not only talk about environmental conditions. Agro-climatic conditions are of course very important for agriculture but I take also into account sociocultural traditions and the way in which people cook and eat. The Soul of Sustainability is also compatibility with human culture, human needs. and human requirements.

Many years ago when I was young and enthusiastic I tried to talk to farmers at Along in Arunachal Pradesh. The farmer was listening very patiently to what I said: "You are growing such a low-yielding variety—why do you stick to it? You can replace it with my variety that will give you about five times more." He just laughed. He said: "I'm growing this variety to make a wine. You demonstrate that it will make the same wine, then I will take your variety!" He said, "I have to use that wine to make an offering to my God. I have to offer this wine to the Sun. Now will your variety do it or not?" I said, "I'm sorry; I don't know!"

Reconciling People and Nature

The harmony between people and nature has been disrupted by modern development. In the past they would cultivate land for a number of years. They didn't come back to that land for fifty-sixty years. By that time it was regenerated and the forest had come back. Today with increasing population pressures those kinds of sustainable measures are not possible anymore. There is no more new land, and therefore we must think of methods of harmonizing nature with human beings in different ways.

We have to develop new methods of management of pests and new methods of replenishment of soil fertility, integrated nutrient supply, bio-fertilizers, organic manures, crop rotations, and legume-cereal rotation.

> By reducing yields, declining land quality can have the same negative effects as declining quantity. Worldwide, according to a 1990 UN assessment, 38 percent of cultivated area has been damaged to some degree by agricultural mismanagement since 1950, with higher levels of degradation in Latin America and Africa. Severe and prolonged land degradation, including soil erosion, nutrient depletion, and desertification, will ultimately remove land from cultivation. Various sources suggest that present losses range from 5 million to 12 million hectares per year.
> —Brian Halweil, *Vital Signs 1999*, p. 42

There is a large amount of land that we call wasteland. Nearly one third of our geographical area is degraded in some way or other. Water harvesting is very important, water saving and sharing. That is an area where you require cooperative management of resources if you want to improve productivity and overcome the constraints. There are inspired leaders who have brought people together to do just that. I believe that we can eradicate hunger, in spite of demographic pressures. There is no doubt at all. In India I believe that the population will stabilize itself at about 1.3 to 1.4 billion by 2050.

Modern and Traditional Methods

Today we can create an integrated package of traditional and modern methods. I don't say that mineral fertilizer can be dispensed with. Suppose I cultivate rice and I require 5 tons of rice per hectare. The minimum nutrients needed by that rice plant is then 100 kilograms of nitrogen. I can give that through urea, a mineral fertilizer. Or I can apply 20–30 kilograms of urea, 20–30 kilograms of bio-fertilizers, and 20–30 kilograms of green manure. There are stem nodulating green manure crops from Senegal called Sesbania rostrata that fix 50 kilograms of nitrogen in 45 days. In this way we can introduce a package that is more eco-friendly.

We cannot dispense with high yields. I would be fooling everybody if I said that I can go back to the past. In the past people were starving. In 1891 the Indian subcontinent had a population of 281 million. Between 1870 and 1900, 30 million children women and men died of hunger. We must neither glorify the past nor abhor the future. We can blend traditional wisdom and modern technology into what we now call eco-technology.

In India we have over 100 million acres of rice. The average yield of rice in India is 2.4–2.5 tons per hectare. Even today China has 5 tons per hectare of yield. In other words, the untapped yield reservoir is still very high, even at current levels of technology. In many countries in Africa it is even greater.

The Famine of Livelihood Opportunities

The major problem is the famine of jobs, of livelihood opportunities. The more people you have, the greater is the problem for them to earn a living. However the present trade structure in the world is very discriminatory against small-scale enterprises. Fortunately the World Bank and leading bilateral donors all stress microcredit, small-scale, and family enterprises.

There is a decisive difference between developing and developed countries. In India 50 percent of our population is below the age of twenty-one. They have to have jobs. We need to fight the famine of jobs and livelihood. I call this a livelihood because a job is a very Western urban concept where you go to work at a particular hour to earn money. Here they have to earn from small-scale enterprises, consisting of one cow here, one buffalo there, a fishpond, a spinning wheel. They have to have multiple enterprises during the day.

In industrialized countries there is a negative population growth and therefore a need for high automation and mechanization. Efficiency means downsizing so they want to remove jobs. In our country that would be suicide. If nothing happens the social disintegration and frustration will become much greater. A recent publication from the Peace Research Institute in Oslo clearly shows that a very large number of conflicts in the last ten years have roots in economic and social disparities.

Orphans in Science

Many scientists are concerned that inequity in the world is growing in social terms and in terms of gender. However science alone is not a magic wand with which we can overcome social problems. There is a very strong viewpoint that science and public policy must become two sides of the same coin.

Research for the public good must continue to be supported from public funds. An increase of intellectual property rights and secrecy in terms of scientific communication is not going to be of long-term benefit to science. There will be problems of plagiarism, misconduct, and jealousy. It is frightening to many senior scientists who until now have experienced a world with free communication and free exchange of material.

In India there are still 1.5 million lepers and leprosy is now a disease of poverty. No company is going to invest in a vaccine because the lepers are too poor. Here you have this problem of orphans remaining orphans in science. By the way, even the market is dependent on much applied work that has been done with public funds in universities and in national laboratories.

Sharing Science

Ninety-two percent of Internet users are from industrialized countries; that signifies a very big gap. The brain drain increases. How do we overcome these enormous disparities in scientific capabilities in the world? You can transfer concepts and technology in some respects, but things like agriculture require much more location-specific technology. I have found from my experience at the International Rice Research Institute that the countries that benefited most from the work of an international center were those that had a strong national research capability. They knew what to collect, what to select, what to apply, and what not to apply.

We need major investments in basic research because that is not an area in which the commercial world is going to invest. Science is the mother of technology; that should not dry up. After the Vienna Conference in 1979 I was invited to chair the UN Advisory Committee on Science and Technology for Development. In 1982 we suggested how the Vienna plan of action must be implemented. To my great disappointment there were no funds. When I was President of the International Union for the Conservation of Nature everybody would suggest ideas, but I would say, "Ladies and gentlemen, conservation without resources is conversation. We can go on conversing but we can't do anything."

Intellectual Property Rights and the Web of Life

How many people know about UNESCO's *Universal Declaration on the Human Genome and Human Rights?* We must have a similar one on the plant genome and farmers' rights. Then bio piracy will give way to bio partnership. Ultimately the web of life is interconnected. Plant, animal, man and woman; we are all connected. We must think of ourselves as part of the web of life, not the master of the earth, the master of the universe, but we are one part of it, living in harmony with the rest of them, then things are in place.

Proactive Mechanisms

All natural resources ultimately have seeds of conflict. Too many people are expecting diverse things out of them and their interests are different so we must harmonize them.

The Narmada Dam in India has elicited enormous passions on both sides. There are people in states like Gujarat who feel that Narmada is their savior; that it's going to result in the irrigation of new areas and so on. There are those who are concerned with human rights issues; there are those who will be displaced by the dam, the tribal families, and the others who will be displaced.

Would there be a United States or an Australia if people had not wanted to migrate from their original location? People are not averse to being taken from one area to another. It depends upon whether they have done it voluntarily, upon how you have done it, what kind of education you have been given, and what's at stake. There is no free lunch; there is always a trade-off.

Could these people who are to be shifted have been involved in the whole process from the beginning? The local people should do the trade-off themselves, not by bureaucrats sitting somewhere. I would say it is a means to an end. In my country Mahatma Gandhi said, "However noble the end may be, the means are equally important." That has been forgotten in administration many times. They think this is good for the people. Often the means adopted undermine the very end itself.

Science and Democracy

How do scientists look upon democratic systems as a method of much greater interaction between science and society? The Swiss referendum on genetically modified organisms is a good example. Many people were surprised that in all the cantons, they voted for the continuation of genetically modified experiments. The environmentalists were very disappointed; they thought it was going to be the reverse. The Swiss Academy President says that the reason is that for the first time scientists went to the streets and started explaining to the common people what genetic modification is, its strengths, risks, and benefits.

> Farmers in the United States sowed their first transgenic crops in 1994, followed by farmers in Argentina, Australia, Canada, and Mexico in 1996. By 1998, 9 nations were growing transgenics, and that number is expected to reach 20–25 by 2000.
> —Brian Halweil, *Vital Signs 1999*, p. 122

Many academies have taken the lead. The Royal Society of London was one of the first to establish the COPUS Committee on the Public Understanding of

Science. They also have a very good journal that the U.S. National Academy is imitating. I'm sorry to say that in developing countries this has not happened yet. Often academies are controlled by government scientists and they are bound by conduct rules. That in itself in my view, is wrong. No academy President should be a government employee.

Patents

When I first went to the United States I was sponsored by the University of Wisconsin Alumni Research Foundation. They had the money to do that because many researchers had assigned their patents to them. Patent money can be used for public good. It is not in itself a bad thing to honor inventors and give them or their companies an incentive for invention.

Another matter is analytic patenting, where you take a medicinal herb and extract the active substance. Here you must have equity in sharing benefits. Article 15 of the Convention on Biological Diversity says that there must be prior informed consent, equity, and ethics in benefit sharing. There are other ethical considerations in terms of patenting because today there is product patenting, process patenting, and product by process patenting.

There is a gold rush today in patenting. I'm sure in another ten years there will be more sanity; more principles and ethics and equity will pervade. Legal regulations are important, but the scientific community must develop a voluntary code of conduct. There should be more credit sharing; more broad-based recognition of a discovery; more equity, and finally, economic benefit sharing.

The Children for Happiness Movement

A very large number of children are being condemned to inadequate brain development at birth due to maternal and fetal malnutrition. The more I see the benefits of modern information technology and the capability of people to absorb knowledge, the more this makes me very sad.

I have been a member of the UN Commission of Nutrition and in our report we have explained in great length the implications of low birth weight, induced by maternal and fetal undernutrition and malnutrition. In my country every third child is of low birth weight, and in Bangladesh it's every second child.

> The principle source of premature death in the world is one of longer standing—malnutrition. The world today is plagued with two nutrition problems: the more traditional undernutrition, and a growing share of the world's population who are overnourished and overweight. The effects of these two forms of malnutri-

tion are essentially the same: increased susceptibility to illness, reduced life expectancy, and lowered productivity. The toll from undernutrition is concentrated in infants and children, whereas nutrition finds its victims later in life. Thus undernutrition has a much greater effect on life expectancy.

—Lester R. Brown, *Vital Signs 1999*, p. 23

When children are born for happiness and not for mere existence, population stabilizes itself. We have seen this in the state of Kerala in India and elsewhere. There is enough grain, the World Food Program has enough grain, the local countries have enough grain, and so on. An interventional mechanism can easily achieve this. There is a lot of wastage of grain and milk powder in Europe with its substantial milk and butter. It can be done and it should be done together with the local village communities. It is partly an educational program, because one has the duty to explain to people the various implications of low birth weight before asking them to join hands with us to overcome this problem.

Once grain area per person drops below a certain level, a nation can lose its ability to feed itself. In Japan, South Korea, and Taiwan, as grain area per person plummeted and incomes soared, grain imports as a share of total consumption have soared from 20 to 70 percent since 1960. In Pakistan, Ethiopia, Iran, and other nations where area per person is already a fraction of the world average, projected population doublings or triplings do not bode well for food security.

—Brian Halweil, *Vital Signs 1999*, p. 42

The cruelest form of inequity is when you are denying the child the opportunity for full development of her innate genetic potential for mental and physical development. We must give this the highest priority.

BRUCE ALBERTS

Sharing Knowledge for Development

Dr. Bruce Alberts, President of the National Academy of Sciences in Washington, D.C., and Chair of the National Research Council, is known for his work both in biochemistry and molecular biology, in particular for his extensive study of the protein complexes that allow chromosomes to be replicated. Dr. Alberts received a doctorate from Harvard University in 1965. He joined the faculty of Princeton University in 1966 and after ten years moved to the Department of Biochemistry and Biophysics at the University of California, at San Francisco, where he became Chair.

He is a principal author of *Molecular Biology of the Cell* (1983), through three editions the leading advanced textbook in this important field. His most recent text, *Essential Cell Biology* (1998), is intended to present this subject matter to a wider audience. He has long been committed to the improvement of science education, dedicating much of his time to educational projects such as City Science, a program that seeks to improve science teaching in San Francisco elementary schools.

Dr. Alberts, as President of the National Academy of Sciences in the United States, what are some of your main concerns—and how did you come to embrace those concerns?

I went to a public high school in the United States and was fascinated by chemistry, in part because in my chemistry class we had bottles of concentrated acids right in front of us that we could use to dissolve metals or cause explosions. This made the science of chemistry come alive. Later, in college I tried to see what I could *do* with chemistry. The only profession that appeared logical at the time was medicine, because doctors had informed me that chemistry was important to their profession. So, I was headed towards becoming a medical doctor, when something happened: In the middle of my college career, I had the great fortune of working for a summer in a research laboratory. Before that time I had taken a lot of science courses, but they didn't teach me what science really was, and so I had kept plodding on toward medical school.

Working that summer in a laboratory completely changed my point of view, and I decided then that I'd try to be a scientist instead of a doctor. Once in graduate school, I became fascinated with the question of how chromosomes replicate, which at that time was a complete mystery. As soon as I became an independent researcher, at the age of twenty-seven, I decided to work on this problem. I spent the next thirty years struggling to work out the puzzle of DNA replication using chemical tools. Throughout these years the advances were so striking that, looking back, we were all amazed to see how fast biology had changed.

Science is an exciting enterprise that builds on knowledge in unexpected ways, and it has been an extremely rewarding occupation for me. There are always new challenges and mysteries to work out. Of course, there is also frustration in science. But maybe once every three years we would make a startling discovery in my laboratory. Doing science is much like solving puzzles. I can't imagine a better occupation.

One of My Great Days in Science

When I was a young professor, I had been working with a mysterious protein that we knew was required to replicate the DNA in chromosomes, but we had no idea what it did. We purified the protein and had it in a test tube. Then one day I mixed it with a sample of DNA. All of a sudden the two strands of the DNA helix came apart in the test tube. The result made me realize quite suddenly that this protein was designed to help open the DNA helix so that its two strands could be replicated. That discovery, made in 1969, actually gave me some stature as a scientist since I soon got tenure at Princeton University, so I owe a lot to that protein.

Originally we thought there was only one protein that replicated DNA, but the process turns out to be much more complicated. Actually the minimal number of proteins required is seven. What I just described was the second one, a single-strand DNA binding protein. The DNA replication apparatus is a lot like a sewing machine that runs along the DNA helix and copies it, both strands at the same time. It's an amazing solution to the problem of chromosome replication. It's much more sophisticated than any of us ever thought, and it was incredibly exciting to actually discover this.

Science for the Twenty-first Century

The *World Conference on Science for the Twenty-first Century* is about two things: Science and its discoveries, and the fact that we need to support science in a way that allows both it and society to prosper in the twenty-first century. We need support for young people, freedom of investigation, and encouragement to try out new ideas. The conference is also about what science means in a New World that's so heavily dependent on science and technology.

The future of the world is all about knowledge. Both in a small country in Africa and in the United States, the future depends not on any natural resource, but rather on the ability to use the world's store of knowledge. In this New World that is so strongly driven by science and technology, knowledge is almost a free good. It's everywhere. Countries are competing on their abilities to use and then build on new knowledge.

The New World Order, which we hope will be based on economic competition and not on military competition, requires that every nation think carefully about how we educate our young people. How do we train them to be able to use and develop scientific knowledge?

This is a great enterprise where everybody wins, because the more we can help other countries develop their ability to generate more knowledge, the more knowledge there will be for us to use. There's just no end to the positive changes that we could bring to our world, if we can all learn to exploit science and technology in beneficial ways.

Sharing Knowledge for Development

As a scientist in the United States for nearly forty years, I was pretty ignorant of the opportunities and challenges for science in addressing the needs of seven-eighths of the world's future population—the people who are growing up now in developing countries. But because of my new job as Academy President, I got out of my laboratory and visited rural farmers' fields in West Africa. I also went to parts of rural India to see what science and technology could do for those people.

The great opportunity before us comes from having new communications devices that can connect everyone. Knowledge can be shared instantaneously and freely over the Internet. Our first challenge is to connect all people to the Internet, and the second challenge is to develop the kind of knowledge resources that will make that connection useful for them. My academy is looking forward to working with developing countries to try some small experiments, because I don't think we know yet how to do this well.

For example, take a small country that has limited scientific ability and like all other countries has the need to purify its water. What are the best technologies? What is the science that can be applied to the problem? How can residents of that nation—any nation—use their water resources with wisdom, so that they have the most effective water supply for the money? All of these things have been done before. How can we share what we know with scientists everywhere, so that the appropriate knowledge can be applied, given how critical these issues are for all societies?

Information Villages in India

Dr. M. S. Swaminathan is a distinguished Indian agricultural scientist and a Foreign Associate of our academy. In recent years he has become a close friend of mine. He has developed something that he calls the information village. It's an experiment to connect twenty or thirty small rural, Indian villages to the Internet using mostly local women as equipment operators. These people are supported by their village to be operators of a local Internet system that connects them to the world's knowledge resources. Volunteers are recruited to ask the village exactly what information they want. Most popular are women's health information, information about local crops (how to grow them best, how to protect them from diseases), daily market prices, and weather forecasts.

This is a wonderful way of bringing the world of knowledge and science into the lives of even the most remote citizens of this planet. These are the types of efforts that we really need to support in the twenty-first century. Otherwise, we're going to have a world of mass poverty. We may even have a major problem of feeding the world in the future. For example, in Africa with its rapidly increasing population, soil degradation is causing less agricultural products to be produced per capita. We will spend much more money later trying to fix something that is broken, than we would spend if we made wise investments now in trying to prevent future disasters.

Americans Want to Help Out

The Consultative Group on International Agricultural Research (CGIAR) has been around for about thirty years. It consists today of sixteen laboratories from all over the world, such as the Philippines Rice Institute and the Mexican Corn and Wheat Institute. The total investment in such public research to reduce poverty needs to be greatly increased. I can't speak for any other country, but if U.S. citizens really understood what the needs of the rest of the world are, and understood what the opportunities are to share knowledge and use our scientific capability to improve the condition of people, I am confident that they would support much larger budgets for this purpose.

Part of my job as President of the U.S. National Academy of Sciences is to create a better communication network between the developing world and our citizens. I know that many young U.S. scientists would love to contribute to meeting world needs. But we need to get systems established so they can find out what the specific challenges are. We also need to have ways of funding such research. We need to convince our people and our government that this is a great long-term investment for a more sustainable world. In fact, their grandchildren are going to benefit directly from living in a more rational and balanced society.

The United States is a very charitable nation. There are large numbers of people who spend their life volunteering, especially after they retire. People in the United States would be very supportive of this kind of investment if they recognized what it would do, and how inexpensive it is compared to major relief efforts or wars.

Patenting versus Sharing of Knowledge

Patents have a finite lifetime of up to twenty years. We need an international system to act as a broker between the industrialized and developed world, if we are to bring important intellectual property resources to the developing world. For example, methods of plant biotechnology should be brought to the developing world without the kind of patent requirements that would be required of an industrialized country. Again, it's in our own interest. People in rural Africa can't pay for the kinds of products that major companies are selling anyway. They are not going to lose markets. Some major U.S. companies are already starting to give away some of their bio-engineered seeds.

I participated in a major report last year that we developed for the World Bank on the international agricultural system, in which this was one of the suggestions. There is an international agricultural research system that has been around already for thirty years, the CGIAR that I mentioned previously. We would like them to step up to the plate and be more aggressive in acting as mediators between the needs of the developing world and the resources of the industrialized world. Their real function is to be a public sector bridge between these two entities.

Bringing Science to the People

In the United States, despite the fact that we are a so-called advanced nation and have a very flourishing scientific and engineering enterprise, there's a big gap between those people who really understand science and engineering and the general public. It's a gap that we need to close, and the National

Academy of Sciences is working very hard with teachers and schools to bring meaningful science into the school curriculum.

Today we have science in most of our schools where the kids fill out a sheet to name twenty different kinds of whales or thirty different kinds of plants, or write down the parts of a plant or a cell. That's really not science. Along with several other organizations in 1996 we developed the first voluntary National Science Education Standards for the United States. These call for a different kind of science. Science should become a core subject starting in kindergarten: hands-on science. "Every child a scientist"; that's our motto. We have developed some wonderful science curricula, but many more teachers have to learn how to use them.

> Surveys and tests regularly show abysmal ignorance about fundamental concepts, principles and facts related to science and technology. One survey showed that 42% of Canadian adults are ill-equipped to work in an increasingly knowledge-intensive environment. Both the USA and Canada fall behind some European countries in "civic scientific literacy". The well-documented chronic failings of many U.S. elementary and secondary schools suggest that there will be no immediate solution to this problem.
>
> —*UNESCO World Science Report 1998*, p. 46

Most teachers have never done science themselves, which is a fault of our university system. Even when you give them good curricula, they need a lot of help in order to teach the children how to do these small science experiments. We work on teacher preparation, spreading good curricula, and on making parents understand that they should look at their schools for a different kind of science. They should expect science that teaches their kid how to think and how to be an effective citizen in the next century.

My Greatest Hope

One of the major outcomes that we as a U.S. delegation at the World Conference on Science would like to accomplish is to help scientists from developing countries convince their own policy makers that science and scientists are important for their country.

The major point is that, in order for a developing country to join the world economy successfully, it must have its own knowledge adapters. You can't be a knowledge adapter unless you're actually experienced with knowledge generation. You therefore need scientists, but scientists doing the kind of science that makes sense for each country. If countries fail to support and nurture their scientific establishment, their universities and their own research efforts, they're going to be left out in the twenty-first century. That is our major message for the world.

References

Alberts, Bruce, et al. 1983. *Molecular Biology of the Cell.* New York: Garland Publishing.
———. 1998. *Essential Cell Biology.* New York: Garland Publishing.

MOHAMED H. A. HASSAN

Let Our Children Bridge the Gap

Mohamed H. A. Hassan

Dr. Mohamed H. A. Hassan was born and educated in the Sudan. Trained as a theoretical plasma physicist, Dr. Hassan is former Professor of Mathematics at the University of Khartoum and Dean of the Faculty of Mathematics. Since 1986 he has served as Executive Director of the Third World Academy of Sciences (TWAS). In April 1999, he was elected President of the African Academy of Sciences. His most recent research interests focus on the physics of wind erosion and sand transport.

Many observers point to Africa's economic and social turmoil and express despair at what they see as daunting problems embedded in centuries of tribalism, slavery, and colonial rule. I do not minimize the depth of Africa's current problems. However, I am hopeful that the continent can regain its economic and social footing by building a sound foundation in science and technology. Debt relief and economic assistance from the North will help. But Africa itself will ultimately play the leading role in its resurgence by drawing on the strong political will of its most outstanding leaders; the desire of all Africans, regardless of where they live, to see their homelands succeed; and the continent's often-forgotten strong commitment to education, which in the 1970s made several African universities among the best in the developing world. My dream is both global and personal: To see my two daughters (and thousands of other youthful students of African-origin who are currently being educated in the West) return home to use their knowledge and skills to help foster an African renaissance.

Dr. Hassan, African science gets remarkably meager press. What we hear instead is the litany of woes; one crisis after another; and the seeming lack of money or interest being focused upon African nations. Could you describe some aspects of scientific research and practice in Africa from your experience?

Few people realize that in the late 1960s and early 1970s, science in Africa was in good shape. In fact, Africa was doing better in some areas of science and technology than South Korea. Some African universities—for example, the University of Makari and the University of Khartoum—enjoyed international reputations for excellence.

The steep slide began in the late 1970s. The reasons were not only political but economic. Financial hardship hit the entire African continent and, unfortunately, institutions of learning and knowledge, including institutions of science, were the first to be adversely affected. The decline was quickly followed by a massive emigration of talent from Africa. The brain drain, which many people think dates back a century or more, actually started just a few decades ago. Before then, education in general and science in particular were in good shape in Africa.

Viewed from the angle of world science, Africa has no achievements to speak of. A 1992 study estimated that Africa counted some 20,000 scientists and engineers, or 0.36% of the world total, and expended 0.4% of the world total on R&D. According to another study, the region was responsible for only 0.8% of total world scientific publications. Its world share of patents is close to zero. Nor has an African won a Nobel Prize for any of the scientific disciplines.

—UNESCO World Science Report 1998, p. 177

The Effects of Globalization

Today, the problems confronting Africa are immense and are of great concern to the rest of the world. Poverty, poor public health, and a deteriorating environment are now global, not just local, problems.

> Recent studies point to the existence of rich indigenous scientific knowledge in many parts of the region. It is claimed that there are written texts which are appreciated as models of environmental management. The region's *Materia Medica* of more than 1,000 animal, plant and mineral products for the treatment of illness is a resource that Western-trained scientists are widely studying. Examples are given of Africa's traditional scientific knowledge in psychiatry, traditional medicine, entomology, trypanosomiasis, agrometeorology, plant production and range ecology and multiple land-use systems. The challenge seems to be how to integrate this African indigenous knowledge (which is seen as holistic and comprehensive, in contrast to the reductionism of Western science) into mainstream analytical science. Many scholars now believe that it will be possible to alleviate many of Africa's endemic problems only by the fusion of modern science and indigenous African knowledge.
>
> *—UNESCO World Science Report 1998,* pp. 177–178

For example, diseases affecting developing countries, especially those in Africa—AIDS and malaria, in particular—have become more widespread. People can travel from one part of the world to another quite easily. Whatever happens today in the Sudan, in Tanzania, or in Ethiopia, the world will know about—and perhaps experience—tomorrow. Globalization, by definition, affects people wherever they are. Consequently, there's worldwide concern about many issues and problems.

The Problem of Debt

Here's a specific example of how the debt problem in Africa weighs on everything, including science. I know that many politicians in the Sudan are interested in the development of science and technology. The minister of science will tell you that he's hopeful. He wants and expects to do something positive, but how can he succeed if the country is financially crippled? The Sudan and countries in similar straits cannot move forward because of long-standing economic handicaps.

> Africa is in dire need of an explicit plan of action, a blueprint that will chart the course of research into the twenty-first century and beyond. Africa consists

mostly of countries that are cash-strapped; many can no longer perform the rudimentary functions of governance; political instability appears pervasive; there are serious threats to law, order and security of life and property; the debt burden of many countries has reached unsustainable levels; available evidence shows that Africa has declined in terms of attractiveness for foreign direct investment; the state of tertiary education has badly deteriorated in most countries; the physical infrastructure of countries is poor and endemic corruption is reported in many.

—UNESCO *World Science Report 1998*, p. 178

Tanzania offers another example of the same hope and the same difficulty. I was speaking to its president who, as a former minister of science, is one of Africa's greatest advocates of science. He's concerned about science for development and about investing more in science. Tanzania invests less than 0.2 percent of its gross domestic product (GDP) in scientific research and development. He wants to increase that figure to 1 percent. His main problem, as he told me, is how to reach that goal if over 60 percent of his country's foreign earnings goes into servicing the debt?

As far as Africa is concerned, advances in science and technology are not being held back by a lack of political will. I'm sure that many African countries, even the poorest ones, possess a political will that is strong enough to push for scientific research and development and for science-based growth. But they are crippled financially. Unless we solve the debt problem, I can't see any way forward.

Capacity Building

Solving the debt problem is not just a question of forgiving debt or imposing austerity measures. It's about capacity building. A few days ago the ministers of science and technology of Africa met and unanimously agreed to present a "debt for science" resolution before the delegates of the World Conference on Science for the Twenty-first Century. They want to transform the debt issue into a science-capacity building issue and to turn this problem into an opportunity for building institutions and developing indigenous capacities. This is the first time a large number of Africa's ministers of science and technology unanimously agreed to such a strategy.

The success of the initiative will ultimately require careful negotiations between the governments and the organizations of the North and international lending organizations, including the World Bank. Each country faces different circumstances and each has different priorities. As a result, debt relief and the obligations it will entail should be done on a case-by-case basis.

The Kind of Science That We Need to Develop

Unfortunately, most areas of science in Africa are in bad shape. However, I would single out the basic sciences—physics, mathematics, biology, and chemistry—for special consideration. These sciences have been hit the hardest over the past few decades. When I served as dean of the faculty of mathematics during the mid 1980s, there were thirty-five Sudanese mathematicians with doctorate degrees. Today, the country only has six Ph.D.s in mathematics. The others have left because of severe difficulties with the economy and extraordinarily poor working conditions.

Many other countries in Africa are in the same shape. The basic sciences are essential for any understanding of science. If a nation wants to do research in agriculture or in health, it must have a critical mass of highly qualified scientists. The same is true when a nation seeks to evaluate the potential impact of a development project.

Support from Nordic Countries

Fortunately, organizations like the Swedish International Development Agency (SIDA) are addressing this issue in a positive way. The agency wants to improve universities and assist in training skilled personnel in the South. Even though SIDA has displayed a consistently strong will to help, its efforts won't be sufficient unless the developing countries themselves take a genuine interest in developing their own capacities in the basic sciences. The hope is that lasting partnerships will be forged between institutions like SIDA in the North and governments in the South. The situation is much better in the agricultural, ecological, and medical sciences than it is in other sciences. But even among the sciences with the best track record, the number of qualified researchers remains below critical mass.

Nordic countries are far more concerned about developing countries than other nations in the North. It's the only region of the world that spends at least 1 percent of its GDP on development assistance. Other rich nations—including France, the United Kingdom, and the United States—spend far lower percentages of their GDPs.

The Basic Sciences

Some countries, both in the North and South, are not convinced that basic sciences are important. They want to invest directly in agriculture and health, seeking immediate solutions to their most pressing problems. I feel that a nation—even a developing nation—must devote a portion of its resources to long-term solutions, which cannot be addressed through applied science alone. There must be room for the basic sciences to grow.

In fact, unless you develop the basic sciences in the countries where the problems are, these problems will never be resolved. That's not to say that you should shut the door and develop your own science. Science is universal, it's developed everywhere, but the South must share in the development of knowledge itself. That's why I call for a partnership in which scientists from the North and the South collaborate on basic science issues of common concern.

The Brain Drain

The biggest and most serious problem in Africa is the brain drain. I hope that delegates to the *World Conference on Science for the Twenty-first Century* formally call for a major effort to be made to build international centers of excellence in several developing countries, especially in the poorest ones. This is the best and most effective mechanisms of counteracting the brain drain.

Countries in the South will then be able to build their capacities locally—and maintain and sustain them—because there's a center where scientists can do world-class research. Salaries hopefully will be sufficient for them to survive. Of course, there will be dramatic discrepancies between what researchers are accustomed to earning in London or Baltimore versus Khartoum. However, once a certain level of material comfort is attained, I think researchers will be more concerned about laboratory and classroom conditions than income. With good laboratories and libraries, you can connect easily to the rest of the world. I think that makes a world of difference.

Present Opportunities

In Khartoum, I've been personally involved in setting up an international center for desert research to deal with the desertification problem. We've built a small laboratory to provide safety measurements related to wind erosion and sand movement. This is one area where the Sudan can be effective in providing suitable field work. It's a huge country with many problems related to desertification. Likewise, a country like Ethiopia could create a center for tropical diseases or for natural products chemistry. Ethiopia has also developed good research capacities in chemistry, which could further evolve into an international center likely to attract talent from all over the world.

The North-South Knowledge Gap

The knowledge gap is huge. Current statistics show that over 90 percent of basic scientific knowledge is produced by northern countries, while less than 10 percent is produced by the southern countries, which represent 80 percent of humanity.

Investment ratios in research and development are the same. The North spends about 90 percent of the world's current total expenditures on research and development. The South contributes about 10 percent. What is even more disturbing is that the gap is widening. As scientific discovery and technological know-how accelerates, many developing countries cannot make sufficient investments in these advances because of more pressing problems or burdensome debts. As a result, they fall even further behind.

Many global problems—issues related to food security, poverty, declining water quality, and expanding deserts—will grow even worse if the knowledge gap between the North and South continues to grow. Such trends, in turn, will have a dire effect on the environment and public health worldwide. The biggest challenge is how to devise mechanisms for narrowing this knowledge gap, which I view as essential prerequisites for solving such chronic problems over the long term.

What the Knowledge Gap Really Means

Food, water, health, and education are the real issues affecting people. People expect their governments to address these issues and governments are searching for experts to devise and implement effective solutions. Scientists should not only be aware of the critical issues facing their societies but should lend their expertise to deal with them.

Just giving more money to poor people will not erase poverty. Proposed solutions must be more sustainable. People who are suffering should be educated so that they can learn how to address their problems. It's partly the responsibility of the scientists not only to provide solutions to problems, but to encourage participation from the community in efforts to improve the everyday lives of people.

Scientists in the developing world are aware of their responsibilities, but the conditions are not right for them to maximize the contributions that they can make to their societies. First, as mentioned earlier, most developing countries have few scientists. Second, scientists who continue to live and work in their home countries face a host of frustrations and must pay a great deal of attention to just handling their academic affairs if they hope to succeed. That often leaves little time or energy for examining the real problems affecting the people. The entire atmosphere must be improved. That raises the issue of governmental policies. Where do the sciences fit in the national agenda? Are the sciences an essential part of larger national plans and aspirations? Just as society has a right to ask what it expects from scientists, scientists have a right to ask what society is likely to provide them with so that they can do the job expected of them.

Science in Developing Countries Is More Complex

Scientists in developing countries face a much more complicated agenda than scientists in developed countries. A chemist working at Massachusetts Institute of Technology often doesn't care about what's happening elsewhere. He or she just wants to do research and get the results published. That's it. But a chemist working in Addis Ababa University is likely to have a larger agenda. He or she is often asked to think about how research or laboratory results might be put to use to address real-life problems facing the community. Much is demanded of scientists in developing countries. They have to face these challenges, but we have to provide them with the right conditions to give them a chance to succeed. The truth is that it's extremely difficult to be a successful scientist in a developing country.

Is There Hope?

The *World Conference on Science for the Twenty-first Century* promises to be a unique event, bringing together leading scientists, ministers of science, and decision makers. Many are likely to return home with a strong message: "Without investing more in science and technology, nothing will improve." I hope that in the long run this creates true North-South partnerships in which both parties feel responsible for narrowing the knowledge gap. Education shortfalls outrun all of the other serious problems likely to be discussed at the meeting. The good news is that the new information technologies could help bridge the knowledge gap between the North and the South.

That's just one sign of hope, but I think this hope is not going to be immediately fulfilled. A great deal of political and economic action is required from all parties. Otherwise, we will just go back to business as usual and things will get worse and worse.

My Message for the Youth, for the Twenty-first Century

Today's youth must be better educated and acquire a better understanding of science. In particular, they must become more involved in understanding real-life problems. They should not only be concerned about producing science, but also about solving real problems. I hope that their governments, especially those in the South, will create the right conditions for them, so that they can undertake this huge challenge.

One of my daughters is studying genetic engineering at MIT. Her dream is to go back to the Sudan and assist in solving real-life problems. My hope is that she will do that. My other daughter is studying economics and political science at Swarthmore College. She's very fond of the United Nations System and hopes to work in a field that will help the UN and other like-minded international institutions achieve their lofty goals.

MARGARET SOMERVILLE

Doing Science in Ethics Time

Margaret Somerville

Australian born and educated Margaret Somerville trained in Pharmacy (Adelaide) and Law (Sydney) and was awarded a Doctorate in Law by McGill University. Currently Somerville is Gale Professor of Law and Professor of Medicine, McGill University, Montreal, and Founding Director of the McGill Centre for Medicine, Ethics, and Law. She plays an active role in the worldwide development of bioethics and the study of the wider legal and ethical aspects of medicine and science. She received several honorary doctorates and the Order of Australia in 1989, and was elected Fellow of the Royal Society of Canada in 1991.

The McGill Centre for Medicine, Ethics, and Law runs programs in Medicine, Ethics, and Law and the Contemporary Canadian Family; programs to Investigate Ethical, Legal, Social, and Economic Impact of HIV/AIDS in Canada; programs in Environment, Health, Ethics, and programs in Law; and programs in Psychiatry, Ethics and Law.

She is an articulate advocate of "doing science in ethics time" rather than "doing ethics after science time," and is a consultant to governments and nongovernmental bodies worldwide.

Professor Somerville, there has perhaps never been so much talk of ethics and science as today. Why is that, and what are some of the leading issues that spur ethical debate in the scientific arena?

There are two test cases: human cloning and xeno transplantation (animal to human transplants). Those two are like the tips of the icebergs that we're going to have to deal with.

There's a wonderful cartoon by Herman in which you see two penguins on an iceberg. They're dressed in little suits with little suitcases in their hands. The iceberg floats into Antarctica and their families are on the shore lined up waiting to meet them. The two penguins on the iceberg say "We're very sorry we're late, but our iceberg hit a ship."

We view the world through very much a lens of our own basic presumptions. We think the ship hit the iceberg because we're on the ship and we see it from our perspective. But we have to always see the reverse presumption that those little penguins are on the iceberg and as far as they were concerned, the ship got in the way of their iceberg. One of the things that we're not doing at the moment adequately, is trying to stand back and look at alternative basic presumptions, so that we can analyze some of the difficult issues that we're facing.

The Ethics of Human Cloning

Human cloning really goes to the essence of what it means to be human. How do we find meaning in human life? What is our relationship to past and future generations? What are the concepts that we need to guard that very fragile reality that human life is? Human life is something wondrous that we have to have the most profound respect for.

For example, we've now got the power to alter the essence of human life itself and how we transmit human life. Therefore we have to develop rules for the way in which we treat human life itself and its transmission. I believe we have to be extraordinarily careful.

At the moment there is in the literature a great deal of concern about potential physical risks to future generations through cloning. But we also have huge, what one might describe as metaphysical risks here, risks to the metaphysical reality we need to live in in order to lead fully human lives—to protect and experience our human spirit.

Risks to the Metaphysical Reality That We Need

By metaphysics I mean, the intangible, invisible, immeasurable reality that we need to help us to find meaning in human life. What will help you live with hope rather than despair? What would help you to be in a positive relationship with other people? I use that term for the collection of values that most people used to find and protect through religion. We no longer have a dominant religion in society; we live in multicultural, pluralistic, often secular societies.

We have to find this sense of deep respect for life that we are prepared to maintain. Out of that comes probably the most important ethical question that we will ask: What should we not do that we now can do, because to do it would be inherently wrong? For me one of those things is human cloning.

Relinquishing Our Power

There's a belief in East Indian culture that the strongest person is the one who when attacked doesn't attack back, but turns and walks calmly away. That is an expression of deciding not to do something. My birthplace is Australia and the Australian aborigines have this extraordinarily powerful belief that they don't own the land; rather the land owns them. They feel that they have this powerful obligation to protect that land, to keep it whole. I believe that life owns us as well as our having life.

Now what about if we felt that way about our own genome? That we can hurt it just the same as we can hurt the land and that we must hold both on trust. The power to do that doesn't make it right to do it. We're the first generation out of millions of years of the evolution that resulted in us, the human species who've actually got this extraordinary power.

We Share All Life in Us

The human genome is the genetic blueprint that goes to make up every human and each of us have got only very slight variations from the other. We humans share 98 percent of our genome with the chimpanzees and 90 percent with a mouse. We've got ancient plant genes in the mitochondrial DNA. We share all life in us, in every cell of our body. Certainly, the human genome has been altered in the past, but that was only through natural processes and natural mutation.

False Analogies and Argument through Confusion

It's very easy to manipulate arguments in discussions about these ethical issues. What often happens is what I call arguing through confusion. Giving something a different name to make it sound as though we're doing something else or not threatening any important values or anything else that people care about and want to protect.

For example, many people think that it is morally wrong to deliberately destroy a human embryo by doing research on it. One way to get around that is to say, I'm doing this research not really on a human embryo; it's a pre-embryo. A rather arbitrary line has been drawn, called a marker event, at fourteen days after fertilization. Before that we're going to call it a pre-embryo and we can do our research. Marker events stop what we do to the embryo becoming a precedent that the same thing could be done to the rest of us.

Reproductive and Therapeutic Human Cloning

More recently, some people are trying to justify what is called therapeutic human cloning. They're dividing human cloning into two streams: one is called reproductive human cloning. That is, you would like to have a little baby that's identical to you, although you might have decided that you would like the baby to have bright blue eyes rather than brown ones.

The other stream is human therapeutic cloning. They will take a cell from your body and make an embryo clone of that cell. Then they'll take one of the stem cells out of that embryo clone. Let's say you need a new liver. They will cause this cell to differentiate in such a way that it makes a new liver for you. It will be genetically identical so you haven't got any of the rejection problems.

One eminent scientist in Britain recently called that stem cell a human tissue generator. Here we are using language to define away ethical problems or moral intuitions. Who would have an objection to a human tissue generator? It's fabulous. You need a liver; you're going to die without it. You want your human tissue generator for your new liver. However, that involves destroying a human embryo because in taking that cell the embryo that is used must be destroyed.

Is it OK to do that? Does it make a difference that we call it human therapeutic cloning? Almost everybody says that human reproductive cloning shouldn't be done. There's not a universal agreement but a lot of people think that's not on. But is it OK to do this other kind of cloning, for therapeutic rather than reproductive purposes? That is the big argument at the moment.

The Shadow of Cloning

Here is this dying person and we could fix him up by doing this and it seems as though you're only using cells; why would you prevent that? It's incredibly hard to speak against doing good.

There is an old saying in human rights: "Nowhere are human rights more threatened than when we act purporting to do good." We're so focused on the good that we fail to see what the shadow of that is. The shadow of doing this human therapeutic cloning is that you are manipulating the very essence of human life and the way in which it is transmitted in the most fundamental way. You are opening up a precedent that says, whatever else we might want to do and how we might want to redesign human life perhaps that is also acceptable.

I was in California arguing with some very avant-garde scientists who think that there's no problem doing this. In fact, they even have a religious argument to support it. They say that so far we've evolved through evolution and that was the divine plan, but the divine plan was that we would find out how to take over evolution and the next stage of human evolution will be humans designing humans.

A Sense of Wonder and Awe

I'd like to pick up on the word wonder because we have to re-find our sense of what I would call wonder and awe. I don't think we can expect to do that in a traditional religious sense and certainly not as a global community. However I think that as the human race we have to do that.

We've got these enormous powers that we can do things that nature never contemplated. I don't think that's just different in degree. I think that's different in kind from anything that we've had before. I would call it argument by confusion, to say that this is just the next step in human evolution and that yes, look at what we've done over the last sixty thousand years and this is just the next thing that we're going to do.

You can look at this extraordinary increase in knowledge in two ways. One is to say, we've got the power to do these things and we're going to do them. The other is to say what we have discovered is extraordinary. You can take the incredible simplicity of the fundamental units that make up the genome and you look at the extraordinary complexity of what has come out of that. That can resurrect our sense of wonder.

The Japanese have a saying that as the radius of knowledge expands, the circumference of ignorance increases. There is a mystery of the unknown that we've got to have the most profound respect for as well as being in wonder and awe at what we do know. We have to balance our knowing with a recognition of our deep unknowing. We need to regard our new knowledge that as requiring what I would call a secular sacred attitude. We have it not just for ourselves, but we must hold it on trust for future generations.

The Rights of the Next Generation

We're intensely individualistic and we make claims on the basis of rights. These rights include someone else's responsibility, but we've been much less forthcoming in articulating those responsibilities. One of those is, of course, what about our responsibilities to the future?

There's a useful concept called memes. They are the accompaniment to genes. Genes are the units of physical information that we pass from generation to generation. Memes are the units of deep cultural information that are passed from generation to generation. The shared story that we tell each other to show that we all belong in the group that shares that story. It is made up of values, principles, attitudes, beliefs, and myths. We equally need to recognize that we can damage or even wipe out those just as we can damage our physical reality. In deciding what we will and won't do with our new capabilities, we will form some of the most important parts of that shared story or wipe them out.

The Future of Human Cloning

Probably therapeutic cloning will be allowed in some countries. The United States is likely to allow it and so is Canada. Canada currently has a bill before Parliament to prohibit cloning, but when you read it, it really only prohibits reproductive cloning because it requires that you do something to the embryo with the intent of forming a child. If you haven't got that intent it wouldn't be caught by the legislation.

That same bill has a provision that would prohibit any alteration of the human germ cell line, the genes we pass on from generation to generation.

Strangely, although I think we should not do embryo cloning, I'm not so sure whether we should totally prohibit germ line alteration. For instance, we could take the cystic fibrosis gene out and put a good one in. I think we would want to do that.

There are several dangers, however. First of all, we don't know what the larger effect of that would be in terms of physical risks for that child. The other danger is that when you open up the acceptability of altering the human germ cell line, you've set a precedent that there's nothing inherently wrong with that. You've opened up the possibility of doing any sort of genetic enhancement, whether intelligence, height, or anything else that somebody wants.

Science and Ethics

There's a rhetoric that uses the word ethics but without a lot of content—ethics is good public relations or even advertising today. The first stage in getting ethics into something is to have people recognize that there is such a thing. In the past, when people came up against an ethical problem in science, most often they identified it as a communications problem: the public just doesn't understand what we're talking about. Or, alternatively, it is considered to be a public relations problem. Many journalists have this experience—a scientist says I can't talk to you about that because our public relations department is handling it.

Ethics is beginning to be seen as yet another hurdle that science has to jump to get on with doing science. Others are becoming more sensitive to the great power that they are discovering and making available and the need to "do ethics" in using this.

The Process of Doing Ethics: Incompressible Time

Doing ethics is a complex process of using all human ways of knowing. Science primarily markets itself as doing something whose end results are always justified by reason.

Ways of human knowing include not just reason, but moral intuition, imagination, creativity, human memory (history), and common sense. I look into what I call examined emotions. You try to examine your emotional response for the ethical knowledge that it can give you. Finally ethics itself is a way of knowing.

Those ways of knowing take a certain time and it may be incompressible time, which is very interesting. Most of what we're doing in science is compressing the time that we need to do things. But in ethics, you can imagine it as a process of sedimentation of values. You have to put all these things in, all these ways of knowing, of feeling, and of imagining. You have to structure them and then you have to let it sit and see what comes out.

You also have to involve a very large number of people because this doesn't just affect the scientist or even just the one person who needs medical treatment. This is something for all of humanity to be involved in deciding. We know that you can't do ethics through a normal democratic process. A majority vote doesn't mean it's ethical. If that were the case, the Nazis were ethical.

How are we going to do this? We ethicists are not the wizards who can come in and fix this all up for you. We can sound warnings; open up structures for people to use. We can identify ethical principles and dilemmas and concepts and ethical tools. But ultimately, there has to be a decision made. That decision has to be taken in some way by all of us and increasingly, the all of us is global.

One of my theories is that some of the response that we need, for instance, what to do about human cloning, will be found as much through poetry as it will be through cognitive analysis. It will be language, structures, metaphors, associations, and feelings as well as reason and intuition that will help us decide what to do. Ethics is a process, not an event.

Doing Science in Ethics Time

In the session on ethics at the *World Conference on Science* I was the only ethicist who spoke. The other people spoke about issues that raised ethical concerns, but that's different from doing ethics.

Everybody needs to talk about ethics. In the same way as when they need to talk about science, they may need the help of people who are professionally trained, or at least have the expertise to guide the doing of the ethics. I think it was very important to get people to start to use the word *ethics* and in the *Declaration on Science* the word ethics takes central stage. This gives us the ethics container, but we haven't got the content yet. The big task is to build the ethics content.

Most people, when they first encounter ethics, particularly scientists, see it as being like something of an add-on. But ethics has to be imbedded in the science. I call it "doing science in ethics time," not doing ethics in science time. Unethical science is bad science no matter how much you discover in doing it. Good science has good ethics.

Bibliography

Sources of Quotes

State of the World 1998
Lester Brown et al. *State of the World 1998*. World Watch Institute. London: Earthscan Publications Ltd. www.worldwatch.org

Vital Signs 1999
Lester Brown et al. *Vital Signs 1999*. World Watch Institute. New York: W. W. Norton & Company.
World Development Report—Knowledge for Development. 1999. The World Bank. Oxford Oxford University Press.

World Science Report 1998
Moore, Howard (ed.). 1998. *World Science Report 1998*. Paris: UNESCO Publishing.
Cetto, Ana Maria (ed.). 2000. *Proceedings of The World Conference on Science—Science for the Twenty-first Century, A New Commitment*. Paris: UNESCO.

Organizations

The American Association for the Advancement of Science (AAAS), www.aaas.org, www.aaas.org/meetings/1999/index.htm
International Forum of Young Scientists, Budapest, Hungary, 23–24 June 1999, www.unesco.org/science/wcs/youth/young.htm
International Council for Science (ICSU), www.icsu.org
United Nations Education, Science, and Culture Organization (UNESCO), www.unesco.org

UNESCO continues to maintain a web site dedicated to the World Conference on Science, with a newsletter and on-line publications publications at www.unesco.org/science/wcs/. Daily reports on the conference were published by the scientific journal *Nature* at their web site: www.nature.com/wcs/

In June 2001 the web site www.scidev.net was launched, addressing issues of science and development. This web site was sponsored by the scientific journals *Nature* and *Science* in association with The Third World Academy of Sciences.

About the Editors

Michael Tobias was on the faculty at Dartmouth College, both in Environmental Studies and the Humanities; has held the Garey Carruthers Chair of Honors at the University of New Mexico-Albuquerque, and was a Visiting Professor, and Regents' Lecturer at the University of California-Santa Barbara. Author and editor of over 30 books, and the writer, director and/or producer of more than 100 films, Tobias is President of the Dancing Star Foundation, which is devoted to animal welfare, biodiversity conservation, and environmental education. Among Tobias' better known works are: The ten-hour dramatic miniseries, *Voice of the Planet,* the book and film, *World War III—Population and the Biosphere at the End of the Millennium, A Vision of Nature— Traces of the Original World, Rage & Reason, The Soul of Nature, Nature's Keepers: On the Frontlines of the Fight to Save Wildlife in America, A Day in the Life of India, America's Great Parks,* and *Antarctica—The Last Continent.*

Teun Timmers has a master's degree in physics and a Ph.D. in Medical Informatics. Before joining Global Vision Network (GVN) he was a scientific researcher for fifteen years and Associate Professor at the University of Amsterdam. He has taught science to journalists and for many years was involved in developing conferences that presented challenging new thinking in science. His main interests are developing programs that highlight new ideas and initiatives in science and society.

Gill Wright established Global Vision Network in 1996 to create a vehicle for new initiatives in the field of science, human development, entertainment, and education. She has a background in the fields of integrated medicine, human development, international leadership issues, and business ethics. She was Vice President of the Club of Budapest International, an organization of high-profile individuals contributing to enriching our future. She was the Executive Producer of *Life on Earth—A True Civilization,* a documentary on the 1996 *State of the World Forum.* GVN was commissioned by UNESCO and ICSU to conduct a media and public relations campaign for the *World Conference on Science for the Twenty-first Century.*

Index